Geometric Crystallography

Geometric Crystallography

An Axiomatic Introduction to Crystallography

Peter Engel

Crystallography Laboratory, University of Bern, Switzerland

D. REIDEL PUBLISHING COMPANY

A MEMBER OF THE KLUWER ACADEMIC PUBLISHERS GROUP

DORDRECHT / BOSTON / LANCASTER / TOKYO

Library of Congress Cataloging in Publication Data

CIP

Engel, Peter, 1942–
Geometric crystallography.

Bibliography: p.
Includes index.
1. Crystallography, Mathematical. I. Title.
QD911.E58 1986 548′.7 86–17894

ISBN-13: 978-90-277-2341-3 **e-ISBN-13: 978-94-009-4760-3**
DOI: 10. 1007/978-94-009-4760-3

Published by D. Reidel Publishing Company,
P.O. Box 17, 3300 AA Dordrecht, Holland.

Sold and distributed in the U.S.A. and Canada
by Kluwer Academic Publishers,
101 Philip Drive, Assinippi Park, Norwell, MA 02061, U.S.A.

In all other countries, sold and distributed
by Kluwer Academic Publishers Group,
P.O. Box 322, 3300 AH Dordrecht, Holland.

Preface

In the last decade mathematical crystallography has found increasing interest. Siginificant results have been obtained by algebraic, geometric, and group theoretic methods. Also classical crystallography in three-dimensional Euclidean space has been extended to higher dimensions in order to understand better the dimension independent crystallographic properties. The aim of this note is to introduce the reader to the fascinating and rich world of geometric crystallography. The prerequisites for reading it are elementary geometry and topological notations, and basic knowledge of group theory and linear algebra.

Crystallography is geometric by its nature. In many cases, geometric arguments are the most appropriate and can thus best be understood. Thus the geometric point of view is emphasized here. The approach is axiomatic starting from discrete point sets in Euclidean space. Symmetry comes in very soon and plays a central role. Each chapter starts with the necessary definitions and then the subject is treated in two- and three-dimensional space. Subsequent sections give an extension to higher dimensions. Short historical remarks added at the end of the chapters will show the development of the theory. The chapters are mainly self-contained. Frequent cross references, as well as an extended subject index, will help the reader who is only interested in a particular subject.

The author is grateful to many persons who have contributed to this note: First of all to my teacher Werner Nowacki who introduced me into crystallography. To Hans Wondratschek for his teaching me crystallographic orbits and four-dimensional space groups. To Wilhelm Plesken for his exposition on higher dimensional lattices at the Bielefeld symposium in summer 1985. To Hans Debrunner for his tutorial on Dehn's function.

I'm especially indebted to Marjorie Senechal for her help and encouragement and to Hans-Rudolf Gnägi for his critical comments. Needless to say that I take full responsability for any errors and misprints and would be grateful for any notification.

I'm also grateful to the D. Reidel Publishing Company for their offer to publish this note and to Michiel C. ten Raa and Ian Priestnall of D. Reidel Publishing Company for their collaboration. The typescript was performed at the computer center of the University of Berne (BEDAG).

Peter Engel

Table of Contents

1. Basic definitions

1.1. Axioms of geometric crystallography 1
1.2. Euclidean vector space 3
1.3. Rigid motions 5
1.4. Symmetry operations 7
1.5. Classifications 10
1.6. Historical remarks 10

2. Dirichlet domains

2.1. Definition of the Dirichlet domain 13
2.2. Some properties of Dirichlet domains 14
2.3. Dirichlet domain partition 16
2.4. A practical method to calculate Dirichlet domains 17

3. Lattices

3.1. The theorem of Bieberbach 22
3.2. Lattice bases 25
3.3. Orthogonal basis 27
3.4. Lattice planes 32
3.5. Dirichlet parallelotopes 33

4. Reduction of quadratic forms

4.1. Definition of the $\mathbb{Z}$-reduced form 44
4.2. The reduction scheme of Lagrange 45
4.3. The reduction scheme of Seeber 46
4.4. The reduction scheme of Selling 59
4.5. The reduction scheme of Minkowski 62
4.6. Historical remarks 66

5. Crystallographic symmetry operations

5.1. Definitions 68
5.2. Rotations in E^2 70
5.3. Rotations in E^n 72
5.4. Symmetry support 81
5.5. General symmetry operations in E^n 84

6. Crystallographic point groups

6.1. Definitions 89
6.2. Point groups in E^2 93
6.3. Point groups in E^3 94
6.4. Point groups in E^n 102
6.5. Root classes 111
6.6. Isomorphism types of point groups 119
6.7. Historical remarks 119

7. Lattice symmetries

7.1. Definitions 123
7.2. Bravais point groups 124
7.3. Bravais types of lattices 128
7.4. Arithmetic crystal classes 141
7.5. Crystal forms 144
7.6. Historical remarks 149

8. Space groups

8.1. Definitions 151
8.2. Derivation of space groups 153
8.3. Normalizers of symmetry groups 165
8.4. Subgroups of space groups 170
8.5. Crystallographic orbits 183
8.6. Colour groups and colourings 191
8.7. Subperiodic groups 195
8.8. Historical remarks 198

9. Space partitions

9.1. Definitions 201
9.2. Dirichlet domain partitions 209
9.3. Parallelotopes 221
9.4. The regularity condition 225
9.5. Dissections of polytopes 234
9.6. Historical remarks 237

10. Packings of balls

10.1. Definitions 240
10.2. Packings of disks into E^2 243
10.3. Packings of balls into E^3 244
10.4. Lattice packings of balls in E^n 246
10.5. Historical remarks 248

References 249

Subject index 261

1. Basic Definitions

The regular shape of crystals suggests that within a crystal atomic building units, congruent to each other, are regularly arranged. Assuming the crystal to be infinite and the atoms to be points, an infinite discrete point set, called a discontinuum, results which plays an essential role in crystallography. Moreover, such point sets are of great importance in several branches of mathematics and physics. Whereas the existence of a continuum in nature cannot be shown, the discontinuum has an assured position in natural sciences. In this chapter some general properties of discrete point sets will be discussed.

1.1. Axioms of geometric crystallography

We consider a point set X in n-dimensional Euclidean space E^n which fulfils, following Hilbert (Hilb2), three conditions:

1.1. The point set X is discrete, that is, around each point of the set an open ball of fixed radius $r>0$ can be drawn which contains no other point of X.
1.2. Every interstitial ball, that is, every open ball which can be embedded into E^n such that it avoids all points of X, has a radius less than or equal to a fixed finite R.
1.3. The point set X looks the same if seen from every point of X.

The second condition ensures that the points are spread uniformly over the whole space. For example they may not lie all on one side of a hyperplane. This signifies that the number of points within any ball of radius $L > R$ increases with the n-th power of L.

A point set X which fulfils the first two conditions is called a discontinuum or, following Delaunay (Delo4), a (r,R)-system. This more general kind of point set is important in the theory of amorphous mater and of quasicrystals.

Following Sohncke (Sohn2), the third condition can be made more precise if we consider the set of straight line segments drawn from any point of the set X to all the

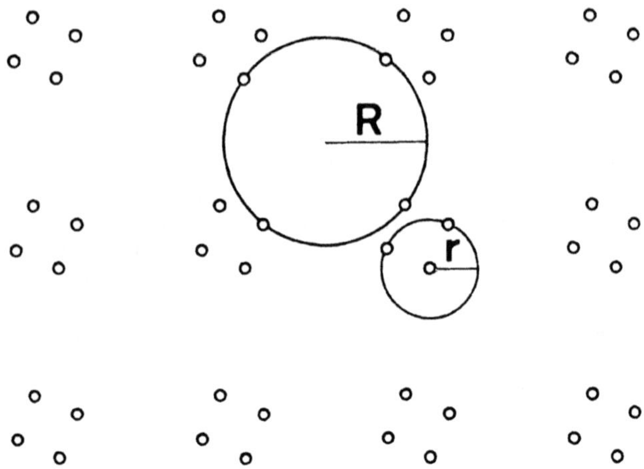

Figure 1.1. A fragment of a regular point system having plane group p4.

remaining points of X. The third condition requires that the line systems of any two points of X are directly or mirror congruent. That is ,for each pair of points we can find a rigid motion of the space which brings the two line systems and hence the whole point set X into self-coincidence.

The third condition ensures that a largest interstitial ball of radius R exists. In a (r,R)-system the radius R is the supreme of radii of all interstitial balls and a ball of radius R not necessarily exists.

A point set X which fulfils all three conditions is called a regular point system by Sohncke (Sohn2) or a homogeneous discontinuum by Niggli (Nigg1).

Regular point systems have applications in the theory of ideal crystals. Any ideal crystal structure can be described as a union of one ore several regular point systems. Each regular point system corresponds to one atomic species.

1.2. Euclidean vector space

We will assume that the reader is familiar with standard linear theory of E^n and elementary topological notations. We also assume familarity with convex sets. The main purpose of this section is to give a brief survey of an Euclidean vector space. The following definitions are standard.

In order to describe the properties of a point set X in n-dimensional Euclidean space E^n, where n is finite, we have to introduce the concept of a real vector space. As origin we choose a point $0 \in E^n$; it need not belong to the point set X. Then we consider the translation which carries 0 to some other point x. This translation can be identified with the vector $\vec{x}$ from the origin 0 to the point x. Selecting n linearly independent vectors $\vec{a}_1,\ldots,\vec{a}_n$ as basis vectors, every vector $\vec{x}$ is uniquely represented by its components $\xi_1,\ldots,\xi_n$ referred to this basis,

$$\vec{x} = \xi_1\vec{a}_1+\ldots+\xi_n\vec{a}_n,$$

The components $\xi_1,\ldots,\xi_n$ can also be considered as the coordinates of the point x.

The dimension n is defined as the maximal number of linearly independent basis vectors.

We represent a vector by a column:

$$\vec{x} := \begin{pmatrix} \xi_1 \\ \cdot \\ \cdot \\ \cdot \\ \xi_n \end{pmatrix} .$$

Defining the sum of two vectors to be

$$\vec{x} + \vec{y} = \begin{pmatrix} \xi_1 \\ \cdot \\ \cdot \\ \cdot \\ \xi_n \end{pmatrix} + \begin{pmatrix} \zeta_1 \\ \cdot \\ \cdot \\ \cdot \\ \zeta_n \end{pmatrix} := \begin{pmatrix} \xi_1+\zeta_n \\ \cdot \\ \cdot \\ \cdot \\ \xi_n+\zeta_n \end{pmatrix}$$

and multiplication by a real scalar λ by

$$\lambda\vec{x} = \lambda \begin{pmatrix} \xi_1 \\ \cdot \\ \cdot \\ \cdot \\ \xi_n \end{pmatrix} := \begin{pmatrix} \lambda\xi_1 \\ \cdot \\ \cdot \\ \cdot \\ \lambda\xi_n \end{pmatrix},$$

a vector space V^n over the field of real numbers is defined.

The vector space V^n is called Euclidean if we define the scalar product of two vectors, referred to the coordinate system $\vec{a}_1,\ldots,\vec{a}_n$, to be

$$\vec{x}\bullet\vec{y} := (\xi_1,\ldots,\xi_n) \begin{pmatrix} c_{11}\ldots c_{1n} \\ \cdot \quad\quad \cdot \\ \cdot \quad\quad \cdot \\ \cdot \quad\quad \cdot \\ c_{n1}\ldots c_{nn} \end{pmatrix} \begin{pmatrix} \zeta_1 \\ \cdot \\ \cdot \\ \cdot \\ \zeta_n \end{pmatrix} = \vec{x}^t C\vec{y},$$

and the length of a vector to be

$$|\vec{x}| := +\sqrt{\vec{x}^t C\vec{x}} \; .$$

The metric tensor $C=(c_{ij})$ is a symmetric matrix with coefficients

$$c_{ij} = |\vec{a}_i|\;|\vec{a}_j|\cos\alpha_{ij},$$

where α_{ij} is the angle between the basis vectors $\vec{a}_i$ and $\vec{a}_j$.

If the basis vectors $\vec{a}_i$ have unit length and are mutually perpendicular, then C is the identity matrix and it follows that

$$\vec{x}\bullet\vec{y} := \xi_1\zeta_1+\xi_2\zeta_2+\ldots\ldots+\xi_n\zeta_n.$$

such a basis is called a cartesian coordinate system.

In crystallography the periodicity of an ideal crystal is used to define a crystal coordinate system which, in general, is not a cartesian one.

Frequently the reciprocal or dual basis $\vec{r}_1,\ldots,\vec{r}_n$ is used. For a vector $\vec{x}:=\xi_1\vec{a}_1+\ldots+\xi_n\vec{a}_n$ and a vector $\vec{y}:=\zeta_1\vec{r}_1+\ldots+\zeta_n\vec{r}_n$ we require that

$$\vec{x}\bullet\vec{y} = \xi_1\zeta_1+\ldots+\xi_n\zeta_n.$$

Thus the reciprocal basis $\vec{r}_1,\ldots,\vec{r}_n$ is obtained by the invers of C, $U:=C^{-1}$.

$$\begin{pmatrix} \vec{r}_1 \\ \cdot \\ \cdot \\ \cdot \\ \vec{r}_n \end{pmatrix} = \begin{pmatrix} u_{11} \ldots\ldots u_{1n} \\ \cdot \qquad \cdot \\ \cdot \qquad \cdot \\ \cdot \qquad \cdot \\ u_{n1} \ldots\ldots u_{nn} \end{pmatrix} \begin{pmatrix} \vec{a}_1 \\ \cdot \\ \cdot \\ \cdot \\ \vec{a}_n \end{pmatrix} ,$$

1.3. Rigid motions

A motion in E^n can be represented by a non-singular nxn matrix S and a shift vector $\vec{s}$; it transforms the coordinates of of a point $x\in E^n$ into those of another point $x'\in E^n$:

$$\begin{pmatrix} \xi'_1 \\ \cdot \\ \cdot \\ \cdot \\ \xi'_n \end{pmatrix} = \begin{pmatrix} s_{11} \ldots\ldots s_{1n} \\ \cdot \qquad \cdot \\ \cdot \qquad \cdot \\ \cdot \qquad \cdot \\ s_{n1} \ldots\ldots s_{nn} \end{pmatrix} \begin{pmatrix} \xi_1 \\ \cdot \\ \cdot \\ \cdot \\ \xi_n \end{pmatrix} + \begin{pmatrix} \sigma_1 \\ \cdot \\ \cdot \\ \cdot \\ \sigma_n \end{pmatrix} .$$

We assume that the point x is moved referred to a fixed coordinate system. Using the Frobenius symbol (Frob1) this equation can be abreviated as

$$x' := (S,\vec{s})\ x.$$

The matrix S is called the rotation part and the shift vector $\vec{s}$ is called the translation part of the motion $(S,\vec{s})$. Every motion that brings x into coincidence with x' has also to bring an arbitrary point $y\in E^n$ into coincidence with some point $y'\in E^n$.

For a rigid motion we require that the length of the vector $\overrightarrow{xy} = \vec{y}-\vec{x}$ is conserved:

$$|\vec{y}-\vec{x}|^2 = (\vec{y}-\vec{x})^t C\ (\vec{y}-\vec{x}) = |\vec{y}'-\vec{x}'|^2$$

$$= |S\vec{y}-S\vec{x}|^2 = (\vec{y}-\vec{x})^t S^t C S\ (\vec{y}-\vec{x})$$

This equation has to remain valid for all $\vec{x},\vec{y}\in E^n$ therefore, $C = S^tCS$. As a necessary and sufficient condition we have

$$c_{ij} = \sum_{h=1}^{n} \sum_{k=1}^{n} c_{hk}\, s_{hi}\, s_{kj}.$$

For a cartesian coordinate system the metric tensor C is the identity matrix and therefore, the following orthogonality relations hold,

$$\sum_{k=1}^{n} s_{ki}s_{kj} = \begin{cases} 0 \text{ if } i \neq j \\ 1 \text{ if } i = j \end{cases}$$

A rigid motion is also called an isometry.

Since det(C) equals $\det(S^tCS)=\det(C)\det^2(S)$ the result $\det(S) = \pm 1$ follows. The two values −1 and 1 for det(S) are connected to the chirality character of the isometry. In oder to understand the chirality character of an isometry $(S,\vec{s})$ we take a subset $M \subset E^n$ of at least n+1 points which not all lie in a hyperplane. In the two-dimensional case we take three points which determine a triangle R in the plane as shown in Figure 1.2. In general n+1 points determine a simplex in E^n. It is always possible to determine a simplex which exhibits chirality, that is, the mirror image of the simplex is not directly congruent to the original simplex. The simplex and its mirror congruent copy are said to be enantiomorph to each other. In Figure 1.2 the triangles R" and R‴ are enantiomorh.

If det(S)=+1 then the isometry $(S,\vec{s})$ carries the simplex into a direct congruent copy. Such an isometry is called a proper isometry. Particularly S is called a proper rotation. If "I" designates the identity operation then $(I,\vec{s})$ is a translation.

Otherwise if det(S)=−1 then $(S,\vec{s})$ carries the simplex into a mirror congruent copy, that is, the chirality of the simplex changes. Such an isometry is called an improper isometry. Particularly S is called an improper rotation or a rotoreflection or, if it leaves a (n−1)-dimensional hyperplane fixed, a reflection.

Example 1.1: A triangle R in the plane E^2 is shown in Figure 1.2. The rotation part S rotates the triangle R into R' through the rotation angle σ around the rota-

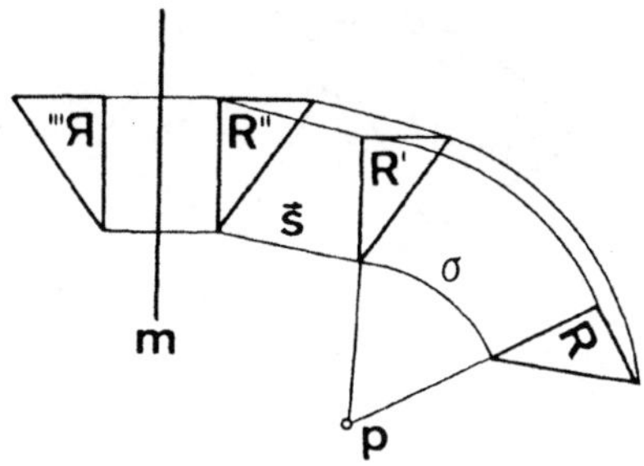

Figure 1.2. Rigid motions in the plane

tion point p. The translation part $\vec{s}$ carries R' into R". Both motions are proper isomteries. The triangle R" is enatiomorph to the triangle R"' hence, there exists no proper isometry in the plane which maps R" onto R"'. However, this can be achieved by a reflection in the mirror line m.

1.4. Symmetry operations

Let M be any subset of E^n. We look at the isometries which map M onto itself.

Definition 1.1: A symmetry operation acting on a set M is an isometry which maps M onto itself.

The symmetry operations of a set M have two important properties:

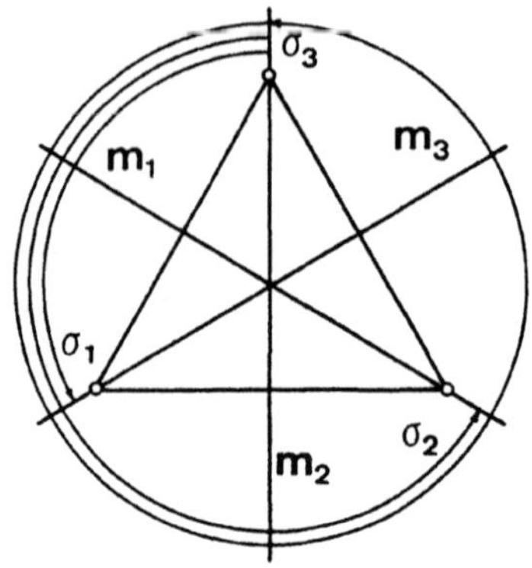

Figure 1.3. Symmetry operations of a set $M \subset E^n$

1. A symmetry operation $(S_1,\vec{s}_1)$ followed by a second symmetry operation $(S_2,\vec{s}_2)$ is again a symmetry operation $(S_3,\vec{s}_3)$ of M.

$$x'' = (S_2,\vec{s}_2)x' = (S_2,\vec{s}_2)(S_1,\vec{s}_1)x$$
$$= S_2S_1x+S_2\vec{s}_1+\vec{s}_2$$
$$= (S_3,\vec{s}_3)x,$$

with

$$S_3 := S_2S_1 \text{ and } \vec{s}_3 := S_2\vec{s}_1 + \vec{s}_2.$$

2. The symmetry operation $(S_3,\vec{s}_3):=(S_1,\vec{s}_1)^{-1}$ which reverses another symmetry operation is again a symmetry operation of M and the result is the identity operation (I,0).

$$(S_3,\vec{s}_3)\ (S_1,\vec{s}_1) = (S_3S_1,S_3\vec{s}_1+ \vec{s}_3) = (I,0).$$

It follows that

$$S_3 = S_1^{-1} \text{ and } \vec{s}_3 = -S_1^{-1}\vec{s}_1.$$

Hence, the totality of symmetry operations of a set M generates a group in the mathematical sense.

Symmetry groups correspond to linear representations of abstract groups in Euclidean vector spaces. Thus we consider $(S,\vec{s})$ as a representation in E^n. We note that different symmetry groups may correspond to different representations of the same abstract group (cf. section 6.6).

Definition 1.2: Every group P of symmetry operations acting on a set M and which leaves at least one point $z \in E^n$ fixed is called a point group.

Example 1.2: Let Δ be the equilateral triangle shown in Figure 1.3. There exist six symmetry operations which map Δ onto itself. These are three rotations S_1, S_2, and S_3 having rotation angles σ_1, σ_2, and σ_3 respectively and three reflections in the mirror lines m_1, m_2, and m_3. The center of gravity of the triangle Δ remains fixed under all these symmetry operations.

We now look at the symmetry operations of a regular point system X. By the regularity condition 1.3 there exists for every pair $x,y \in X$ a symmetry operation $(S,\vec{s})$ which carries x into y and thereby maps X onto itself. It follows that all $x \in X$ are connected through symmetry operations acting on X. If this is fulfilled we say that the group of symmetry operations acts transitively on X.

Definition 1.3: Every group G of symmetry operations acting transitively on a regular point system in E^n is a n-dimensional space group.

Synonymous with space group also crystallographic group is used.

Definition 1.4: The set of all symmetry operations of a group Γ which map a set M onto itself is called the stabilizer of M in Γ.

Synonymous with stabilizer also site symmetry group or isotropy group are used.

1.5. Classifications

We will make frequent use of classifying objects into similarity classes. Such classifications are based on the following definition:

Definition 1.5: A relation, denoted by "~", between elements a,b,... of a set M which is

1. reflective, $a \sim a$,
2. symmetric, $a \sim b \rightarrow b \sim a$,
3. transitive, $a \sim b,\ a \sim c \rightarrow b \sim c$,

is called an equivalence relation.

Each equivalence relation on a set M divides M into disjoint subsets, called equivalence classes. Synonymous with equivalence class also type will be used in order to be in accordance with the nomenclature of the International Tables, Vol. A (Hahn1).

1.6. Historical remarks

The study of discrete point sets is intimately connected with the investigation of the geometrical properties of crystals, which began in the 16th century with Georgius Agricola (1494-1555), Conrad Gessner (1516-1565), Johann Kepler (1571-1630), and Christian Huygens (1629-1695). Remarkably, the ancient Greeks and Romans did not investigate crystal forms. Detailed descriptions of the beginnings of crystallography are contained in the books of Lenz (Lenz1), Marx (Marx1) and Burke (Burk1).

The first important crystal law was discovered by Niels Stenson (1636-1686) in 1669 when he stated the law of constancy of interfacial angles which he had observed for quartz crystals. "ad evincendum, in plano axis laterum et numerum et longitudinem varie mutari, non mutatis angulis" (Sten1). In its full generality this law was asserted only after Jean Baptiste Louis Romé de l'Isle (1736-1790) had measured many different crystal species (Romé1).

Further progress was achieved by René Just Haüy (1743-1822) who explained the crystal forms by stacking of equal parallelepipedal building bricks called "molécules soustractives" which themselves were constructed by "molécules intégrantes".

A modern atomistic theory was introduced in 1824 by the physicist Ludwig August Seeber (1793-1855) who postulated small spherical atoms which remain in a stable equilibrium by the balance of attractive and repulsive forces. These atoms were assumed to be situated at the nodes of three mutually intersecting sets of equidistant parallel planes thus forming a point lattice (Seeb1). Independently in 1843 Gabriel Delafosse (1796-1878) considering only the centers of gravity of the molecules, argued that these centers have to form a point lattice (Dela1).

These ideas of Seeber and Delafosse opend a new area in crystallography: the investigation of discrete point sets. At that time it was assumed that the atoms or molecules in a crystal would have to be arranged parallel in space and thus the mass centers would form a point lattice. An extension of this arbitrary assumption of point lattices was proposed in 1863 by Christian Wiener (Wien1). Taking up the ideas of Wiener in 1867 the physicist Leonhard Sohncke (1842-1897) proposed that an infinite discrete point system is one which shows the same distribution of points about each of its points (Sohn1). Later in 1874 Sohncke gave a rigourous definition (Sohn2):

> <u>A dicrete point system is regular if from any two points of the system straight lines are drawn to all the other points of the system and these two line systems are directly or mirror congruent.</u>

Sohncke explicitly considered direct and mirror congruent line systems. But then he arbitrarily assumed that in nature only direct congruence occurs. This question was settled correctly only after several crystals were described whose shape could only be explained with improper symmetry operations. In 1885 Pierre Curie (1859-1906) gave an example of a crystal form (he did not say what crystal species it was) which could only be generated by a fourfold rotoreflection (Curi1). In 1887 Leonhard Wulff showed that also for the minerals dioptase, $Cu_6(Si_6O_{18})\cdot 6H_2O$, and phenakite $Be_2(SiO_4)$, improper rotations are required (Wulf1).

We notice that Sohncke already used explicitly conditions 1.1 and 1.3 and implicitly he assumed condition 1.2. The complete set of conditions was enumerated by David

Hilbert (1862-1943) in his lecture on intuitive geometry (Hilb2).

Sohncke investigated the regular point systems in the plane using only geometric arguments. For the first time he enumerated the 13 different construction types of regular point systems in the plane which are now called the 13 types of eigensymmetry groups in the plane.

At that time the successfull determination of the space-group types simultaneously by Arthur Schönflies (1853-1928) and by Evgraph Stepanovich von Fedorov (1853-1919) interrupted the geometric investigation of regular point sets (Schö3, Fedo2). These investigations were taken up again by the crystallographer Paul Niggli (1888-1953) and by the mathematician Boris Nikolaevič Delaunay and his school.

Leonhard Sohncke was born in Halle (Germany) in 1842. He studied at Halle and got his theses in 1866. He became a lecturer in Königsberg and in 1871 he was appointed a professor in physics at the Polytechnical Highschool of Karlsruhe. Later he went to Jena and in 1888 he was appointed a professor in physics at the Polytechnical Highschool of Munich where he remainded the rest of his life. He died in 1897.

Evgraph Stepanovich von Fedorov was born in Orenburg (Russia) in 1853. His father was a military engineer. The family moved to St. Petersburg (Leningrad) where the boy visited the German school and then he attended the school of military engineering. Between 1872 and 1874 he was an active officer in the Ukraine. He then resigned and continued his studies at the medical school of St. Petersburg. There, he was imprisoned because of his revolutionary activities against the tsar régime. He managed to escape to Germany and Switzerland. Later in 1880 he continued his studies at the Mining-academy of St. Petersburg. In 1895 he became a professor in mineralogy and geology at the Highschool for agriculture of Moscow and in 1905 he was appointed director of the Mining-academy of St. Petersburg. He died there in 1919.

2. Dirichlet Domains

We continue the investigation of point sets. In this chapter the set X is assumed to be a (r,R)-system. General properties of X will be discussed. More examples and results will be presented in chapter 9.

2.1. Definition of the Dirichlet domain

The Dirichlet domain (after the mathematician Peter Gustav Lejeune Dirichlet) is an important mathematical tool for investigating point sets. The metrical and topological properties of a point set $X \subset E^n$ can best be seen from the Dirichlet domain partition.

Definition 2.1: For $x_0 \in X$ the Dirichlet domain $D(x_0) \subset E^n$ is the part of space containing all points which are closer to x_0 than to any other point $x_i \in X$.

The point x_0 is called the reference point of the Dirichlet domain $D(x_0)$.

Synonymous with Dirichlet domain also Voronoĭ region, Wigner-Seitz cell, Brillouin zone (used in reciprocal space), domain of influence, or plesiohedron are used.

The reference point x_0 is identified with the origin. We take any other point $x_i \in X$ and determine the bisecting hyperplane H_i° which is normal to the straight line segment $\overline{x_0 x_i} = \vec{x}_i$. By construction all points $p \in H_i^\circ$ have equal distance from x_0 and x_i. The hyperplane H_i° separates the space into two open half-spaces H_i^+ and H_i^-. We assume H_i^+ to contain x_0. It follows that all points $p \in H_i^+$ lie closer to x_0 than to x_i as is shown in Figure 2.1.

We determine the open half-space for each point $x_i \in X \setminus x_0$. Thus the open Dirichlet domain $D(x_0)$ is the intersection of all half-spaces H_i^+,

$$D(x_0) = \bigcap_{x_i \in X \setminus x_0} H_i^+ .$$

The half-spaces H_i^+ are convex hence, $D(x_0)$ is convex.

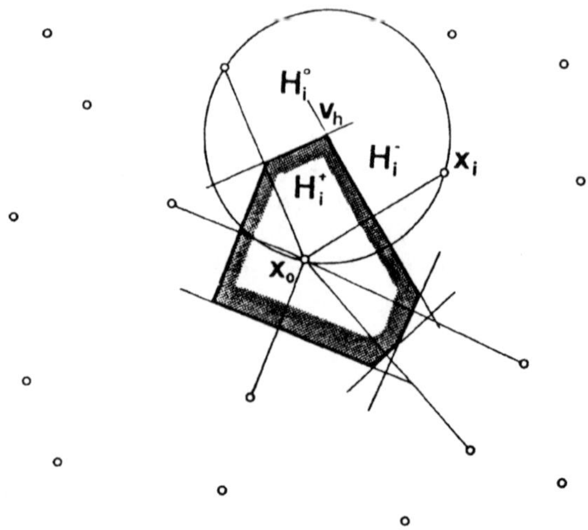

Figure 2.1. Construction of the Dirichlet domain

2.2. Some properties of Dirichlet domains

For any point $p \in D(x_0)$ we consider the open ball $B(p, |\vec{p}|)$ with center in p and radius $|\vec{p}|$. We take such a ball for each point $p \in D(x_0)$ and obtain the region

$$Q_0 = \bigcup_{p \in D(x_0)} B(p, |\vec{p}|).$$

Theorem 2.1: All points $x_i \in X$ which generate facets of $D(x_0)$ lie on the boundary of Q_0.

Proof: Suppose there exists a point $x_j \in X$ inside a ball $B(p, |\vec{p}|)$. Then the point p would be closer to x_j than to x_0. Therefore, p could not belong to $D(x_0)$ which is a contradiction. A point x_j outside of Q_0 is outside of any ball $B(p, |\vec{p}|) \subset Q_0$ hence, for every $p \in D(x_0)$ it follows that $|\overrightarrow{x_j p}| > |\overrightarrow{x_0 p}|$ and therefore the hyperplane H_j^0 bisecting $\overline{x_0 x_j}$ does not intersect $D(x_0)$.

Theorem 2.2: In a (r,R)-system X every Dirichlet domain $D(x_i)$, $x_i \in X$, is a bounded convex polytope with a finite number of boundary facets. $D(x_i)$ is determined through all points within a ball of radius 2R and center x_i.

Proof: By theorem 2.1, all points of X generating facets of $D(x_i)$ lie on the boundary of Q_i. By condition 1.2 the radius R is the supreme of radii of all interstitial balls. Therefore Q_i is contained within a ball $B(x_i, 2R)$. By condition 1.1 there are only a finite number of points of X within the ball $B(x_i, 2R)$.

We denote by $clD(x_i)$ the closure of $D(x_i)$. A point $p \in clD(x_i)$ is called an extreme point if p is not a relative interior point of any closed straight line segment $[p_1, p_2] \subset clD(x_i)$. (For details the reader is referred to Bron1).

For a bounded convex polytope the vertices v_j are the extreme points. If the number of facets is finite then also the number of vertices is finite. The following theorem holds.

Theorem 2.3: In a (r,R)-system X the region Q_i of any Dirichlet domain $D(x_i)$, $x_i \in X$, is determined through the union of all balls at the vertices v_j of $clD(x_i)$,

$$Q_i = \bigcup_{v_j \in clD(x_i)} B(v_j, |\overrightarrow{x_i v_j}|)$$

Proof: For any point $p \in D(x_i)$ there exists a closed straight line segment $[p_1, p_2]$, $p_1, p_2 \in clD(x_i)$, such that $p \in int[p_1, p_2]$. We have that

$$B(p, |\overrightarrow{x_i p}|) \subset [B(p_1, |\overrightarrow{x_i p_1}|) \cup B(p_2, |\overrightarrow{x_i p_2}|)].$$

Every point $p \in D(x_i)$ can be represented as a linear combination of the vertices $v_j \in clD(x_i)$ (cf. section 9.1). From the observation that the vertices v_j are the only points which are not relative interior points of $clD(x_i)$ the above theorem follows.

Theorem 2.4: In a (r,R)-system X the Dirichlet domain $D(x_i)$ of any point $x_i \in X$ is contained within a ball of radius R and center x_i.

Proof: By condition 1.2 the radius R is the supreme of radii of all interstitial balls. All points of $D(x_i)$ lie in the interior of the ball $B(x_i,R)$. Hence $D(x_i) \subset B(x_i,R)$.

2.3. Dirichlet domain partition

We construct the Dirichlet domain $D(x_i)$ for each point $x_i \in X$. This results in a partition of space which has the following properties (cf. section 9.1):

2.1. The union of all closed Dirichlet domains $\mathrm{cl}D(x_i)$ cover the space,

$$\bigcup_{x_i \in X} \mathrm{cl}D(x_i) = E^n.$$

2.2. The intersection of the interior of any two Dirichlet domains is empty,

$$\mathrm{int}D(x_i) \cap \mathrm{int}D(x_j) = \emptyset, \quad i \neq j.$$

2.3. The intersection of two closed Dirichlet domains is either empty or a d-face of each.

The following theorem 2.5 reveals the important covering property of the balls $B(x_i,R)$.

Theorem 2.5: If we attach at each point x_i of a (r,R)-system X a closed ball of radius R then

$$\bigcup_{x_i \in X} \mathrm{cl}B(x_i,R) = E^n.$$

Proof: By construction we have

$$\bigcup_{x_i \in X} \mathrm{cl}D(x_i) = E^n.$$

By theorem 2.4, $\mathrm{cl}D(x_i) \subset \mathrm{cl}B(x_i,R)$ holds. Hence, theorem 2.5 is proved.

2.4. A practical method to calculate Dirichlet domains

A closed Dirichlet domain $clD \subset E^3$ is a convex polyhedron P consisting of 0-, 1-, and 2-dimensional elements called vertices v_h, edges E_i, and faces F_j respectively. Among these elements exists a partial order, $v \subset E \subset F \subset P$, which allows a hierarchical structure (for details the reader is referred to Nef2).

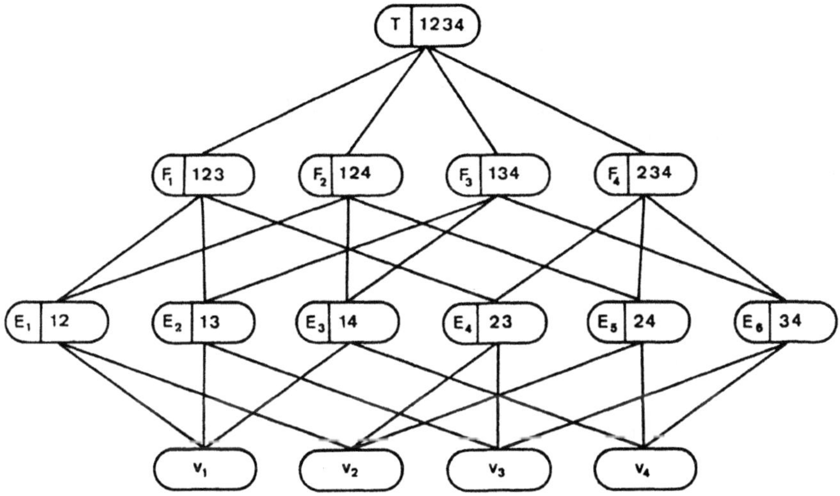

Figure 2.2. Hierarchical structure of a tetrahedron

Figure 2.2 illustrates the hierarchical structure of a polyhedron. The elements of a polyhedron are arranged in different levels. The zero level contains the vertices. The first level contains the edges. Each edge has two subordinate vertices. A subordinate relationship is indicated in the scheme with a line segment connecting an edge E_i with a subordinate vertex v_h. The second level contains the faces. If the face F_i is a k-gon then F_i has k subordinate edges. On the other hand two faces meet in every edge. A vertex is determined by at least three faces therefore, every vertex must be subordinated to at least three faces. Finally the third level element is called the head of the polyhedron P. The head has all subordinate faces.

In a similar way the hierarchical structure of a n-dimensional convex polytope $P \subset E^n$, $n \geq 2$, can be set up. The d-dimensional elements of P are called d-faces. Particularly the 0-faces are called vertices, the 1-faces are called edges, and the (n-1)-faces are called facets. Every d-face F_h is a d-dimensional convex polytope. To each d-face F_h are subordinated all the (d-1)-faces $F'_j \subset F_h$. In Figure 2.3 such subordinate relationships are indicated with line segments connecting the d-faces with the subordinate (d-1)-faces. Again we observe that every 1-face is subordinated by two 0-faces. In general a d-face, d=0,...,n-2, is determined by at least n-d facets therefore, every d-face is subordinated to at least n-d facets. Finally, every (n-2)-face is subordinated to exactly two facets.

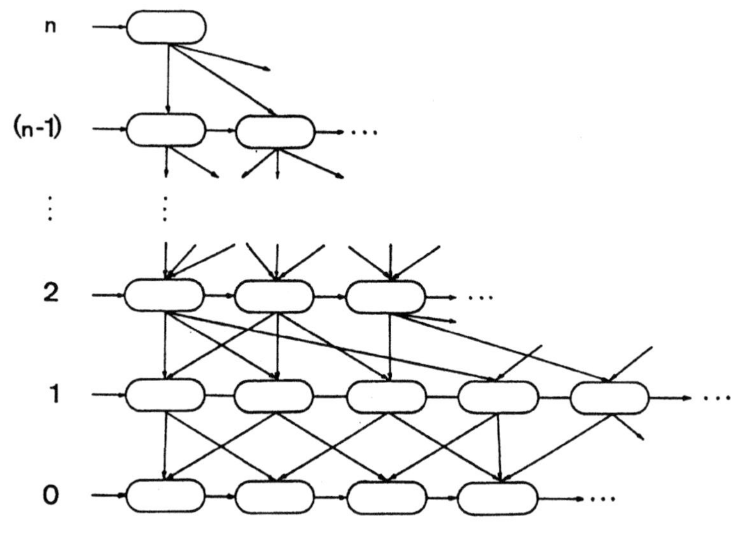

Figure 2.3. Scheme of a hierarchical structure of a polytope

In order to calculate the polytope P we start from a sufficiently large simplex $A_n \subset E^n$ such that $P \subset A_n$. The simplex in E^3 is the tetrahedron. Generally, a simplex in E^n is the convex hull of n+1 points not all of them lying in a hyperplane. The simplex has a particularly simple hierarchical structure which is easily set up. The simplex A_n is intersected with all half-spaces H_i^+, i=1,...,m, which define the polytope P,

$$P = A_n \cap H_1^+ \cap \ldots \cap H_m^+.$$

The calculation of P proceeds along the following steps:

1. We set up the hierarchical structure of a simplex $A_n \subset E^n$. To do this we take n+1 points in E^n and assign them to the n+1 vertices of A_n. In reference to a cartesian coordinate system with origin $0 \in \mathrm{int}P$ we assign for the vertices v_i, i=1,...,n, the coordinates $(0,\ldots,0,-\xi_i,0,\ldots,0)$ and for the vertex v_{n+1} the coordinates $(\xi_1,\ldots,\xi_n)$, where $\xi_i=\xi$ is a real number. For the remaining steps we have to consider only two levels, say d and d-1, d=1,...,n, at a time. A d-face $F_h \subset A_n$ is determined through d+1 points out of the n+1 points. There are $\binom{n+1}{d+1}$ different possibilities to select d+1 points out of n+1 points. Thus there are $\binom{n+1}{d+1}$ different d-faces. Recall that the binominal coefficient,

   $$\left|\begin{matrix} h \\ k \end{matrix}\right| := \frac{h(h-1)\ldots(h-k+1)}{1\ 2\ \ldots\ k},$$

   is defined for integers $h \geq k > 0$. To each d-face F_h we subordinate the adjacent (d-1)-faces. Every (d-1)-face $F'_j \subset F_h$ is determined through d points out of the d+1 points of F_h. There are d+1 possibilities for F'_j which we connect to F_h. After having performed the subordination for all F_h we proceed to the next higher level until we have processed level n. As an example of this step we may look at Figure 2.2. There, the numbers in each box indicate the vertices which are contained in the corresponding d-face. For computational purposes we connect the d-faces F_h within each level d, d=0,...n-1, in order to process every level directly.

2. We take the half-space H_1^+ and determine the half-space intersection $A_n \cap H_1^+$. We assume the half-space to be given by the facet vector $\vec{f}_1$ orthogonal to the hyperplane H_1^0 and by the distance $\delta>0$ of H_1^0 from the origin. The hyperplane H_1^0 divides the space into two half-

spaces H_1^+ and H_1^- respectively, where H_1^+ is assumed to contain the origin. In a first step we calculate for each vertex v_h if it is contained in H_1^+, H_1^o or H_1^- and assign to it the code '001', '010', or '100' respectively. This is decided upon

$$\Delta := \delta - \vec{f}_1 \bullet \vec{v}_h.$$

Because of rounding errors we use a small limit $\varepsilon>0$. The following cases may occur:

$\Delta \geq \varepsilon$	if $v_h \in H_1^+$	code '001'
$abs(\Delta)<\varepsilon$	if $v_h \in H_1^o$	code '010'
$\Delta \leq -\varepsilon$	if $v_h \in H_1^-$	code '100'

If the code is '100' then the corresponding vertex v_h is cut off by H_1^o and therefore, it is bypassed in the 0-level.

The operations which have to be performed on the higher level elements now depend on the operations applied to the lower level elements. In the remaining steps we have to consider at most three levels at a time, say d, d-1, and d-2 (only in the case that d>1), for d=1,...,n. The code of a d-face F_h is the Boolean "or" union of the codes of the subordinate (d-1)-faces. If all (d-1)-faces $F'_j \subset F_h$ lie in H_1^o, or H_1^- that is, the resulting code of F_h is '010', or '110' respectively, then F_h is cut off by H_1^o and therefore, it is bypassed in the d-level. If there exist (d-1)-faces $F'_j \subset F_h$ on both sides of H_1^o that is, the resulting code is '101' or '111' respectively then a new (d-1)-face F'_n is generated by H_1^o. We add F'_n with initial code '010' to the scheme at level d-1 and connect F'_n to F_h.

We now look in detail at each of the (d-1)-faces $F'_j \subset F_h$. If the code of F'_j is '100', or '110' respectively then the corresponding subordination is removed. If the code is '001','011', '101', or '111' then two cases have to be considered. First if d=1 then the intersection of the edge $E_h:=F_h$ with the hyperplane H_1^o has to be calculated, $v_n:=F'_n=E_h \cap H_1^o$. Second if d>1 then we have to consider the (d-2)-faces $F''_k \subset F'_j$. If the code of F''_k is '010' that is, $F''_k \subset H_1^o$, then F''_k has to be connected to the new (d-1)-face F'_n. This process is illustrated in Figure 2.4.

In a similar way we process all $F'_j \subset F_h$. Afterwards we proceed to the next h, h:=h+1, until the level d is finished. Then we proceed to the next level d, d:=d+1, until level n is finished. This way the half-space intersection is completed.

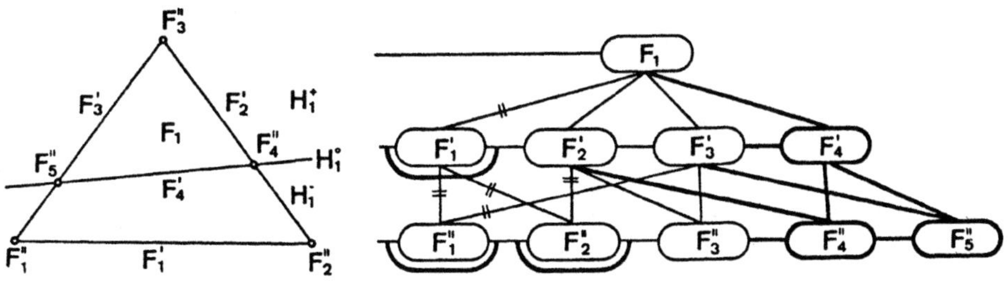

Figure 2.4. Half-space intersection

3. We take the next half-space until all half-space intersections are completed.

For the calculation of a Dirichlet domain $D(x_0)$ of a point $x_0 \in X$ we determine after each half-space intersection the domain

$$Q_0 := \bigcup_{v_h \in \mathrm{cl}D'} B(v_h, |\vec{v}_h|),$$

where $D(x_0) \subset D'$. As soon as there are no further points $x_i \in \mathrm{int}Q_0$ the calculation of $D(x_0)=D'$ is completed.

This routine is very efficient and gives a correct result within the ε-limit also in cases, where many redundant coincidences occur. The procedure is independent of the dimension n, however, even for low dimensions, n=5, or n=6, the number of d-faces may become very large so that a complete calculation of a polytope is beyond any existing computer capabilities. A computer program to calculate Dirichlet domains of regular point systems has been developed according to this procedure (Enge5).

3. Lattices

In this chapter $X \subset E^n$ is assumed to be a regular point system. We investigate the periodic property of X.

3.1. The theorem of Bieberbach

For a given regular point system X we construct the Dirichlet domain partition τ. We choose $x_0 \in X$ as origin. By condition 1.3 and by theorem 2.1, every $D(x_i) \in \tau$ is congruent to $D(x_0)$. By theorem 2.2, $D(x_0)$ has a finite number of facets, hence there exists a point group $P(x_0)$ of finite order which maps τ onto itself and thereby maps $D(x_0)$ onto itself.

$$P(x_0) := \{W_i \mid W_i: \tau \to \tau,\ D(x_0) \to D(x_0)\}.$$

Condition 1.3 requires a space group G which acts transitively on X hence, G also acts transitively on τ, that is, for every pair $D(x_i)$, $D(x_j) \in \tau$ there exists a symmetry operation $(S, \vec{s}) \in G$ which carries $D(x_i)$ onto $D(x_j)$ and thereby maps τ onto itself.

Theorem 3.1: The rotation parts of the symmetry operations $(S, \vec{s}) \in G$ generate a point group of finite order.

Proof: From the observation that $D(x_0)$ has a finite number of facets F_i it follows that there exists a finite set M of symmetry operations $(S_i, \vec{s}_i) \in G$ each of which maps $D(x_0)$ onto a facet sharing $D(x_i)$,

$$M := \{(S_i, \vec{s}_i) \mid (S_i, \vec{s}_i): D(x_0) \to D(x_i),\ \mathrm{cl}D(x_0) \cap \mathrm{cl}D(x_i) = F_i\}.$$

If $(S_i, \vec{s}_i) \in M$ then also $(S_i, \vec{s}_i) W_j \in M$ for every $W_j \in P(x_0)$. For each $D(y_v) \in \tau$ there exists an infinite number of different chains $D(x_0), \ldots, D(y_v)$ of mutually facet sharing Dirichlet domains. For a particular chain $D(x_0), D(y_1), \ldots, D(y_v)$ we determine the symmetry operation $(R_v, \vec{r}_v)$ which maps $D(x_0)$ onto $D(y_v)$ by successively applying along this chain conjugated versions of some $(S_i, \vec{s}_i) \in M$ as is shown in Figure 3.1,

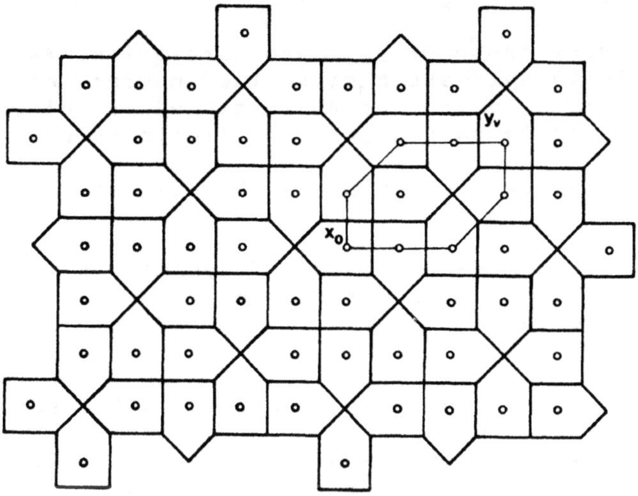

Figure 3.1. A fragment of a Dirichlet domain partition having plane group p4

$$D(y_1) = (S_h,\vec{s}_h)D(x_0) := (R_1,\vec{r}_1)D(x_0)$$
$$D(y_2) = (R_1,\vec{r}_1)(S_j,\vec{s}_j)(R_1,\vec{r}_1)^{-1}D(y_1)$$
$$= (S_h,\vec{s}_h)(S_j,\vec{s}_j)D(x_0) := (R_2,\vec{r}_2)D(x_0)$$
$$\vdots \qquad \vdots$$
$$D(y_v) = (S_h,\vec{s}_h)\ldots(S_k,\vec{s}_k)D(x_0) := (R_v,\vec{r}_v)D(x_0).$$

Because such a chain can be found for every $D(x)\in\tau$ it follows that M is a generating set of G.

For any other chain $D(x_0),\ldots,D(y_v)$ the resulting symmetry operation

$$(R_v,\vec{r}_v) = (S_m,\vec{s}_m)\ldots(S_s,\vec{s}_s)$$

may only differ by some symmetry operation $W_j\in P(x_0)$. For the rotation part we obtain

$$R_v = S_m\ldots S_s = S_h\ldots S_k W_j,$$

from this follows the relation

$$W_j^{-1}S_k^{-1}\ldots S_h^{-1}S_m\ldots S_s = I.$$

For every chain we get such a relation. As M and $P(x_0)$ are finite sets and because the number of facets of $D(x_0)$ is finite and at least n+1 it follows that the $S_i \in M$ generate only a group of finite order.

For further developments the theorem of Bieberbach is fundamental (Bieb1).

Theorem 3.2: Every group of symmetry operations in E^n which acts transitively on a regular point system X contains n linearly independent translations.

Proof: The space group G contains an infinite number of symmetry operations. By theorem 3.1, the rotation parts generate a point group of finite order therefore, at least two different symmerty operations $(S,\vec{s})$, $(S,\vec{r}) \in G$ exist which have the same rotation part S. It follows that

$$(S,\vec{r})^{-1}(S,\vec{s}) = (S^{-1}S,S^{-1}\vec{s}-S^{-1}\vec{r}) = (I,\vec{t}),$$

where $\vec{t}:=S^{-1}\vec{s}-S^{-1}\vec{r}$, is a translation. Let $(I,\vec{t}_1)$ be the smallest such translation. It generates a subgroup of infinite order,

$$T^1 := \{(I,\vec{t}) \mid \vec{t}=m_1\vec{t}_1,\ m_1 \in \mathbb{Z}\}.$$

If we apply T^1 to $D(x_0)$ a linear array of equidistant Dirichlet domains is generated. We determine the right coset $T^1(S,\vec{s})$. It generates a parallel array of Dirichlet domains. The number of such arrays is infinite hence, there exists an infinite number of cosets. It follows that two symmetry operations belonging to different cosets exist which have the same rotation part. As before, a second translation $(I,\vec{t})$ results which is linearly independent of $(I,\vec{t}_1)$. Let $(I,\vec{t}_2)$ be the shortest such translation. The translation vectors $\vec{t}_1$, $\vec{t}_2$ generate a subgroup of infinite order

$$T^2:=\{(I,\vec{t}) \mid \vec{t}=m_1\vec{t}_1+m_2\vec{t}_2,\ m_1,m_2 \in \mathbb{Z}\}.$$

Similar arguments show that n linearly independent translations can be found which proves theorem 3.2.

Bieberbach's theorem reveals that every space group contains a subgroup of infinite order,

$$T^n := \{(I,\vec{t}) \mid \vec{t}=m_1\vec{t}_1+\ldots+m_n\vec{t}_n,\ m_1,\ldots,m_n\in\mathbb{Z}\}.$$

Each coset $T^n(S_i,\vec{s}_i)$ generates a congruent lattice. We represent every lattice vector $\vec{t}\in T^n$ through its endpoint and thus obtain a point lattice. It follows that a regular point system can be represented as a finite set of congruent point lattices, each parallelly placed in space. This is the periodic property we promissed to show.

3.2. Lattice bases

We now investigate in E^n the lattice

$$T^n := \{\vec{t} \mid \vec{t}=m_1\vec{a}_1+\ldots+m_n\vec{a}_n,\ m_i \in \mathbb{Z}\},$$

where $\vec{a}_1,\ldots,\vec{a}_n$ are n linearly independent lattice vectors.

Definition 3.1: A lattice basis in E^n consists of n linearly independent lattice vectors $\vec{a}_1,\ldots,\vec{a}_n$ with the property that each lattice vector $\vec{t}\in T^n$ may be represented as a linear combination $\vec{t}=m_1\vec{a}_1+\ldots+m_n\vec{a}_n$ with integers m_i.

The metric of the lattice with basis vectors $\vec{a}_1,\ldots,\vec{a}_n$ is described by the metric tensor C (cf. section 1.2).

The squared volume of the parallelepiped P spanned by the basis vectors is given by the determinant of C,

$$\mathrm{vol}^2(P) = \det(C).$$

A new lattice basis, $\vec{a}'_1,\ldots,\vec{a}'_n$, is obtained by a linear transformation of the old lattice basis with the transformation matrix $A:=(a_{ij})$ with integral coefficients,

$$\begin{pmatrix} \vec{a}'_1 \\ \cdot \\ \cdot \\ \cdot \\ \vec{a}'_n \end{pmatrix} = \begin{pmatrix} a_{11}\ldots a_{1n} \\ \cdot \qquad \cdot \\ \cdot \qquad \cdot \\ \cdot \qquad \cdot \\ a_{n1}\ldots a_{nn} \end{pmatrix} \begin{pmatrix} \vec{a}_1 \\ \cdot \\ \cdot \\ \cdot \\ \vec{a}_n \end{pmatrix}.$$

In order that $\vec{a}'_1,\dots,\vec{a}'_n$ again is a lattice basis the volume of the new parallelepiped vol(P') has to be equal to vol(P). Hence, the transformation matrix A must be unimodular, that is, det(A)=±1.

The number of such matrices A is unlimited, which means, in E^n, $n>1$, an infinite number of lattice bases exist for a lattice. Among these lattice bases we choose a suitable one as representative with respect to which all calculations may easily be performed. Several reduction procedures will be described in chapter 4 which allow us to find for each lattice a unique representative.

In reference to the lattice basis $\vec{a}_1,\dots,\vec{a}_n$, a lattice vector $\vec{t}$ can be written as

$$\vec{t} = m_1\vec{a}_1+\dots+m_n\vec{a}_n$$

with integral components m_i. The squared length of $\vec{t}$ becomes

$$|\vec{t}|^2 = \vec{t}^{\,t}C\vec{t} = \sum_{i=1}^{n}\sum_{j=1}^{n} c_{ij}\, m_i m_j \quad := f(m_1,\dots.m_n).$$

The function $f(m_1,\dots,m_n)$ is called a positive-definite n-nary quadratic form that is, it is non-negative and assumes zero only for $m_1=\dots=m_n=0$.

Referred to a new lattice basis $\vec{a}'_1,\dots,\vec{a}'_n$, the lattice vector $\vec{t}$ becomes

$$\vec{t}\,' = m'_1\vec{a}'_1+\dots+ m'_n\vec{a}'_n.$$

The new components $m'_1,\dots,m'_n$ transform according to the contragredient matrix $A^o := (A^{-1})^t$,

$$\vec{t}\,' = A^o\vec{t},$$

where A is the transformation matrix of the basis vectors.

We require that the length of the lattice vector $\vec{t}$ is conserved under a transformation of the lattice basis:

$$|\vec{t}|^2 = \vec{t}^{\,t}C\vec{t} = \vec{t}'^{\,t}ACA^t\vec{t}\,' = \vec{t}'^{\,t}C'\vec{t}\,' = |\vec{t}\,'|^2.$$

It follows that the metric tensor C', referred to the new lattice basis $\vec{a}'_1,\dots,\vec{a}'_n$, becomes

$$C' = ACA^t.$$

This is an equivalence relation. The metric tensor C' is said to be equivalent to C. Different lattice bases of one and the same lattice have equivalent metric tensors and the corresponding quadratic forms are also said to be equivalent.

We now calculate the transformation required for an isometry $(S,\vec{s})$ under a change of basis vectors. Referred to the lattice basis $\vec{a}_1,\ldots,\vec{a}_n$, we have

$$y = (S,\vec{s})x.$$

Referred to a new lattice basis $\vec{a}'_1,\ldots,\vec{a}'_n$ obtained through the transformation A, the coordinates of x and y transform according to the contragredient matrix A^o,

$$x' = A^o x, \quad y' = A^o y.$$

It follows that

$$A^t y' = (S,\vec{s})A^t x'$$

and

$$y' = A^o(S,\vec{s})A^t x' = (U,\vec{u})x'.$$

Therefore,

$$(U,\vec{u}) = (A^o S A^t, A^o\vec{s}).$$

3.3. Orthogonal basis

In many cases it is convenient to work with a cartesian coordinate system instead of using a lattice basis. A set of mutual orthogonal basis vectors $\vec{e}'_1,\ldots,\vec{e}'_n$ is obtained from Schmidt's orthogonalization procedure as follows:

$$\begin{aligned}
\vec{e}'_1 &= \vec{a}_1\\
\vec{e}'_2 &= \lambda_{21}\vec{e}'_1 + \vec{a}_2\\
\vec{e}'_3 &= \lambda_{31}\vec{e}'_1 + \lambda_{32}\vec{e}'_2 + \vec{a}_3\\
&\vdots\\
\vec{e}'_n &= \lambda_{n1}\vec{e}'_1 + \ldots + \lambda_{nn-1}\vec{e}'_{n-1} + \vec{a}_n.
\end{aligned}$$

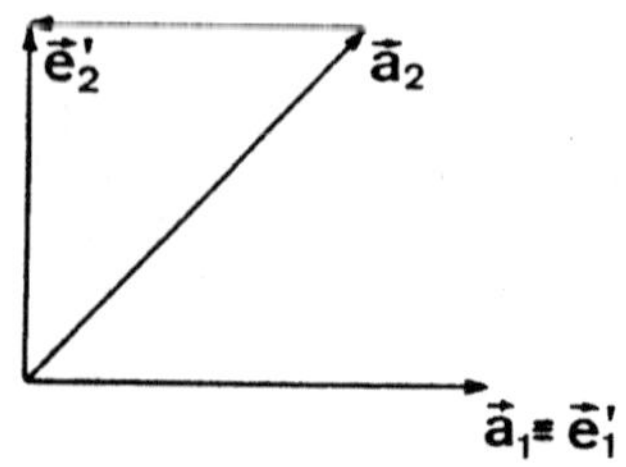

Figure 3.2. Schmidt's orthogonalization procedure

The coefficients λ_{ij} can be determined successively. Assume that $\vec{e}'_1, \ldots, \vec{e}'_{i-1}$ are already determined. From the condition that $\vec{e}'_i \bullet \vec{e}'_j = 0$ for $i \neq j$ we obtain

$$\vec{e}'_j \bullet \vec{e}'_i = 0 = \lambda_{ij}(\vec{e}'_j \bullet \vec{e}'_j) + \vec{e}'_j \bullet \vec{a}_i$$

hence,

$$\lambda_{ij} = -\frac{\vec{e}'_j \bullet \vec{a}_i}{|\vec{e}'_j|^2} .$$

After a normalization a cartesian coordinate system $\vec{e}_1, \ldots, \vec{e}_n$ is obtained

$$\vec{e}_i := \frac{\vec{e}'_i}{|\vec{e}'_i|} .$$

The following practical algorithm is virtually based on the same orthogonalization procedure. But instead of transforming the basis vectors we transform the coordinates starting from the metric tensor C. The principel is to transform the quadratic form

$$f(\xi_1,\ldots,\xi_n) = \vec{x}^t C\vec{x} = \sum_{i=1}^{n}\sum_{j=1}^{n} c_{ij}^{(1)}\xi_i\xi_j$$

into a $\mathbb{R}$-reduced form

$$f(\zeta_1,\ldots,\zeta_n) = \zeta_1^2+\ldots+\zeta_n^2.$$

To start with we set $\vec{y}=C\vec{x}$. The quadratic form then becomes

$$f(\xi_1,\ldots,\xi_n) = \vec{x}^t C\vec{x} = \vec{x}^t\vec{y} = \xi_1\eta_1^{(1)}+\ldots+\xi_n\eta_n^{(1)}.$$

The transformation $\vec{y}=C\vec{x}$ gives the equations

$$\begin{aligned}
\zeta_1' := \eta_1^{(1)} &= c_{11}^{(1)}\xi_1+c_{12}^{(1)}\xi_2+\ldots+c_{1n}^{(1)}\xi_n\\
&\;\;\vdots\\
\eta_n^{(1)} &= c_{n1}^{(1)}\xi_1+c_{n2}^{(1)}\xi_2+\ldots+c_{nn}^{(1)}\xi_n.
\end{aligned}$$

We now replace ξ_1 by $\eta_1^{(1)}$ using the algorithm of Gauss (cf. section 5.4). This is always possible because C is positive-definite. We obtain

$$\begin{aligned}
\xi_1 &= c_{11}^{(2)}\zeta_1'+ c_{12}^{(2)}\xi_2+\ldots+ c_{1n}^{(2)}\xi_n\\
\eta_2^{(1)} &= c_{21}^{(2)}\zeta_1'+ c_{22}^{(2)}\xi_2+\ldots+ c_{2n}^{(2)}\xi_n\\
&\;\;\vdots\\
\eta_n^{(1)} &= c_{n1}^{(2)}\zeta_1'+ c_{n2}^{(2)}\xi_2+\ldots+ c_{nn}^{(2)}\xi_n,
\end{aligned}$$

where $c_{11}^{(2)} := 1/c_{11}^{(1)}$, $c_{1j}^{(2)} := -c_{1j}^{(1)}/c_{11}^{(1)}$, $c_{i1}^{(2)} := c_{i1}^{(1)}/c_{11}^{(1)}$, and

$c_{ij}^{(2)} := c_{ij}^{(1)} - c_{i1}^{(1)}c_{1j}^{(1)}/c_{11}^{(1)}$, $i,j>1$.

The quadratic form then becomes

$$\begin{aligned}
&f(\zeta_1',\xi_2,\ldots,\xi_n)\\
&= (c_{11}^{(2)}\zeta_1'+ c_{12}^{(2)}\xi_2+\ldots+c_{1n}^{(2)}\xi_n)\zeta_1'+\ldots+\xi_n(c_{n1}^{(2)}\zeta_1'+\ldots+ c_{nn}^{(2)}\xi_n)\\
&= c_{11}^{(2)}\zeta_1'^2+ \sum_{i=2}^{n}\sum_{j=2}^{n} c_{ij}^{(2)}\xi_i\xi_j.
\end{aligned}$$

Thus we have split off a quadratic form $f_1(\zeta_1') = c_{11}^{(2)}\zeta_1'^2$. Geometrically this means that the Euclidean vector space

E^n is decomposed into a one-dimensional subspace U^1 and its orthogonal complement U^{n-1}. The remaining quadratic form

$$f(\xi_2,\ldots,\xi_n) = \sum_{i=2}^{n} \sum_{j=2}^{n} c^{(2)}_{ij}\xi_i\xi_j$$

can be split up in the same way. We have

$$\begin{array}{lll} \zeta'_2 := \eta^{(2)}_2 = & c^{(2)}_{22}\xi_2+\ldots+c^{(2)}_{2n}\xi_n \\ \quad\vdots & \quad\vdots \qquad\qquad\quad \vdots \\ \quad\eta^{(2)}_n = & c^{(2)}_{2n}\xi_2+\ldots+c^{(2)}_{nn}\xi_n. \end{array}$$

We exchange $\eta^{(2)}_2$ by ξ_2 and get the quadratic form

$$c^{(2)}_{11}\zeta'^2_1+ c^{(3)}_{22}\zeta'^2_2+ \sum_{i=3}^{n} \sum_{j=3}^{n} c^{(3)}_{ij}\xi_i\xi_j.$$

We continue in this way until we obtain

$$c^{(2)}_{11}\zeta'^2_1+\ldots+c^{(n)}_{nn}\zeta'^2_n.$$

We still have to normalize

$$\zeta_i=\zeta'_i\sqrt{c^{(i+1)}_{ii}} = \zeta'_i/\sqrt{c^{(i)}_{ii}}.$$

Altogether we have the transformation

$$\begin{array}{ll} \zeta_1 = & (c^{(1)}_{11}\xi_1+ c^{(1)}_{12}\xi_2+\ldots+c^{(1)}_{1n}\xi_n)/\sqrt{c^{(1)}_{11}} \\ \zeta_2 = & \qquad\qquad (c^{(2)}_{22}\xi_2+\ldots+c^{(2)}_{2n}\xi_n)/\sqrt{c^{(2)}_{22}} \\ \vdots & \qquad\qquad\qquad \ddots \qquad \vdots \\ \zeta_n = & \qquad\qquad\qquad\qquad\qquad c^{(n)}_{nn}\xi_n \;/\sqrt{c^{(n)}_{nn}}. \end{array}$$

So the coordinates transform with an upper triangular matrix L. The basis vectors transform according to the contragredient matrix $U=(u_{ij}):=L^{\circ}$,

$$\vec{e}_1 = u_{11}\vec{a}_1$$
$$\vec{e}_2 = u_{21}\vec{a}_1 + u_{22}\vec{a}_2$$
$$\vdots$$
$$\vec{e}_n = u_{n1}\vec{a}_1 + u_{n2}\vec{a}_2 + \ldots + u_{nn}\vec{a}_n.$$

Thereby, the connection to Schmidt's orthogonalization procedure is established.

Example 3.1: We take in E^3 the metric tensor

$$C := \begin{vmatrix} 4 & 2 & 2 \\ 2 & 5 & 3 \\ 2 & 3 & 6 \end{vmatrix}$$

and set

$$\zeta'_1 = 4\xi_1 + 2\xi_2 + 2\xi_3.$$

We exchange ξ_1 by ζ'_1 and obtain the new metric tensor in E^2,

$$\begin{vmatrix} 4 & 2 \\ 2 & 5 \end{vmatrix} .$$

We set

$$\zeta'_2 = 4\xi_2 + 2\xi_3.$$

After exchanging ξ_2 by ζ'_2 we get

$$\zeta'_3 = 4\xi_3.$$

After normalization we obtain the upper triangular matrix L for the transformation of the coordinates,

$$L = \begin{vmatrix} 2 & 1 & 1 \\ 0 & 2 & 1 \\ 0 & 0 & 2 \end{vmatrix} .$$

Finally we calculate the contragredient matrix $U := L^\circ$ for the transformation of the basis vectors,

$$U = \begin{vmatrix} \frac{1}{2} & 0 & 0 \\ -\frac{1}{4} & \frac{1}{2} & 0 \\ -\frac{1}{8} & -\frac{1}{4} & \frac{1}{2} \end{vmatrix} .$$

3.4. Lattice planes

We consider a d-dimensional sublattice T^d of T^n.

Definition 3.2: For a lattice vector $\vec{t} \in T^n$ the coset $\vec{t}+T^d$ is a d-dimensional lattice plane.

Particularly if d=1 the coset is called a lattice line and, if d=n-1, it is called a lattice hyperplane or, if n=3, a lattice plane.

A lattice hyperplane F is characterized by the facet vector $\vec{f}$ orthogonal to F. The following theorem is known as the law of rationality:

Theorem 3.3: (law of rationality) Every lattice hyperplane F has an orthogonal facet vector $\Delta\vec{f}$ which has integral components if referred to the reciprocal basis $\vec{r}_1,\ldots,\vec{r}_n$.

Proof: Let $T^{n-1} \subset T^n$ be the sublattice defining the lattice hyperplane F. Assume T^{n-1} is spanned by n-1 linearly independent lattice vectors $\vec{u}_i$ referred to a lattice basis $\vec{a}_1,\ldots,\vec{a}_n$ of T^n,

$$\vec{u}_i = m_{i1}\vec{a}_1+\ldots+ m_{in}\vec{a}_n, \quad i=1,\ldots,n-1, \quad m_{ij} \in \mathbb{Z}.$$

Let C be the metric tensor of the lattice basis $\vec{a}_1,\ldots,\vec{a}_n$. The reciprocal basis $\vec{r}_1,\ldots,\vec{r}_n$ is obtained through the invers matrix C^{-1} (cf. section 1.2). We refer the facet vector $\vec{f}$ to the reciprocal basis,

$$\vec{f} = \mu_1\vec{r}_1+\ldots+ \mu_n\vec{r}_n.$$

Because $\vec{f}$ is orthogonal to T^{n-1} we have

$$\vec{f}\cdot\vec{u}_i = 0 = \mu_1 m_{i1}+\ldots+\mu_n m_{in}, \quad i=1,\ldots,n-1.$$

As the $\vec{u}_i$, i=1,...,n-1, are linearly independent it follows that n-1 of the μ_i can be exchanged by the algorithm of Gauss (cf. section 5.4). Assume that μ_k remains. We choose arbitrarely $\mu_k=1$. The μ_i, $i=1,\ldots,n$, $i\neq k$, then can be calculated

$$\mu_i = q_{ik}\mu_k = q_{ik},$$

where the coefficients q_{ik} are taken from the final exchange scheme. Since the q_{ik} are rational a smallest integer $\Delta>0$ exists such that

$$\Delta\vec{f} = h_1\vec{r}_1+\ldots+ h_n\vec{r}_n$$

with integral components $h_1,\ldots,h_n$.

The integers $(h_1,\ldots,h_n)$ are called the Miller indices of the lattice hyperplane F.

The cosets of T^{n-1} define an infinite stack of lattice hyperplanes F_k. We can label the F_k's in the following way:

$$\ldots, F_{-2}, F_{-1}, F_0=T^{n-1}, F_1, F_2,\ldots$$

For each F_k the number k is given by its representative $\vec{t}$,

$$k:= \vec{t}\bullet\Delta\vec{f}.$$

The lattice hyperplane spacing δ between two adjacent lattice hyperplanes is given by

$$\delta:= 1/|\Delta\vec{f}| = 1/\sqrt{\Delta\vec{f}^{t}C^{-1}\Delta\vec{f}}.$$

In real space crystals grow by addition of atoms, or clusters of atoms, to its surface. The correspondence rule predicts that the new material is added in layers along lattice planes having large lattice plane spacing δ. Thus the polyhedral shape of crystals is mainly determined by such lattice planes.

3.5. Dirichlet parallelotopes

If we take the point x_0 as origin then the action of the lattice

$$T^n:=\{\vec{t} \mid \vec{t}=m_1\vec{a}_1+\ldots+m_n\vec{a}_n, m_i\in\mathbb{Z}\}$$

on x_0 generates a point lattice. Around the origin x_0 we construct the Dirichlet domain $P(x_0)$ (cf. section 2.1). It has the property that translates of $clP(x_0)$ cover Euclidean space E^n.

Definition 3.3: A parallelotope P is a convex polytope congruent copies of which tile Euclidean space E^n by translation.

The Dirichlet domain $P(x_0)$ is a special kind of parallelotope; we call it a Dirichlet parallelotope.

There are two combinatorial types of parallelogons in the plane E^2 which were used since the antique for pavings. These are the centrosymmetric hexagon and the rectangle. An example of a paving with hexagons, which is inspired from the honycombe, is shown in Figure 3.3. An other example of a tessellation with squares is the chessboard.

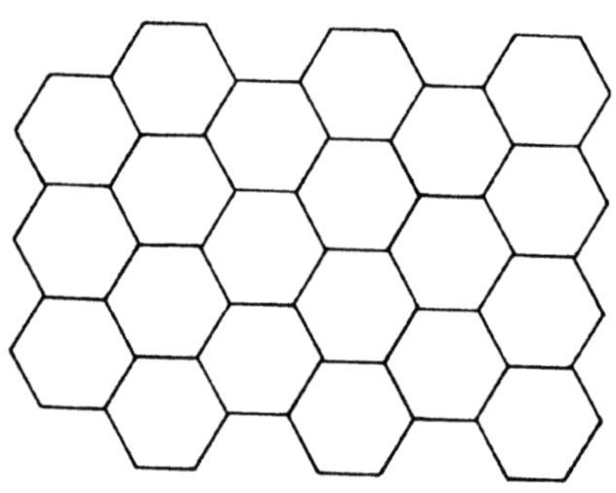

Figure 3.3. Fragment of a paving of the plane E^2 with hexagons having plane group p6mm

In this section we investigate some properties of Dirichlet parallelotopes. More examples and results will be presented in chapter 9. Although we restrict ourselves to Dirichlet parallelotopes most results remain valid for general parallelotopes. Voronoĭ (Voro1) implicitly conjectured that every parallelotope is combinatorially equivalent to a Dirichlet parallelotope. Delaunay (Delo1) proved this conjecture for $n \leq 4$.

<u>Theorem 3.4:</u> The Dirichlet parallelotope $P(x_0)$ is centrosymmetric and each facet F of $P(x_0)$ is centrosymmetric.

Proof: For every lattice vector $\vec{t} \in T^n$ also $-\vec{t} \in T^n$. Therefore T^n has a center of symmetry in x_0 hence, $P(x_0)$ is centrosymmetric. For any lattice vector $\vec{v} \in T^n$ also $\vec{t}-\vec{v} \in T^n$. Therefore, a center of symmetry exists at $\frac{1}{2}(\vec{v}+(\vec{t}-\vec{v})) = \frac{1}{2}\vec{t}$. It follows that each facet is centrosymmetric.

If the lattice vector $\vec{f}$ carries $P(x_0)$ onto $P(x_i)$ such that $\mathrm{cl}P(x_0) \cap \mathrm{cl}P(x_i)=F$ is a facet of $P(x_0)$, then $\vec{f}$ is called a facet vector.

We now give conditions for a lattice vector to be a facet vector. Let T' be the sublattice

$$T' := \{\vec{t} \mid \vec{t}=2m_1\vec{a}_1+\ldots+2m_n\vec{a}_n,\ m_i \in \mathbb{Z}\}$$

of index 2^n. Then T^n can be written as the union of $h:=2^n$ cosets,

$$T^n = T' \cup \vec{t}_2+T' \cup \ldots \cup \vec{t}_h+T'.$$

The following theorem of Voronoĭ (Voro1, No. 48) completely characterizes the facet vectors of $P(x_0)$.

<u>Theorem 3.5:</u> The lattice vectors $\vec{f}$ and $-\vec{f}$ are facet vectors of $P(x_0)$ if and only if $\vec{f}$ and $-\vec{f}$ are exactly the two shortest vectors in their coset $\vec{f}+T'$.

Proof: Let $\vec{f}$ be a facet vector of $P(x_0)$. It follows that $\frac{1}{2}f \in \mathrm{cl}P(x_0)$. For any $\vec{t} \in T^n$, $\vec{t} \neq \lambda \vec{f}$, also $\vec{f}' := \vec{f}+2\vec{t} \in \vec{f}+T'$. Suppose that $|\vec{f}'|<|\vec{f}|$. As shown in Figure 3.4 we take

$$\vec{p} := \tfrac{1}{2}(\vec{f}+\vec{f}') = \tfrac{1}{2}(2\vec{f}+2\vec{t}) = (\vec{f}+\vec{t}) \in T^n.$$

It follows that

$$|\vec{p}-\tfrac{1}{2}\vec{f}| = |\tfrac{1}{2}\vec{f}+\vec{t}| = \tfrac{1}{2}|\vec{f}'| < \tfrac{1}{2}|\vec{f}|$$

hence, $\frac{1}{2}f \notin \mathrm{cl}P(x_0)$ which is a contradiction. This means that the bisecting hyperplane H° of $\vec{p}$ cuts off $\frac{1}{2}f$. This remains true for every $\vec{t} \neq \lambda \vec{f}$ therefore, if $\vec{f}$ and $-\vec{f}$ are

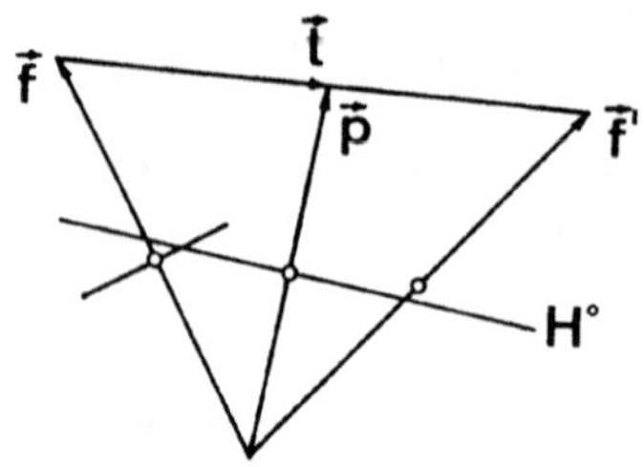

Figure 3.4

the only shortest vectors in their coset $\vec{f}$+T' then they are facet vectors. In the limiting case that $|\vec{f}|=|\vec{f}'|$ then neither $\vec{f}$ nor $\vec{f}'$ is a facet vector.

Un upper limit for the number of facets of $P(x_0)$ is given by the following theorem due to Minkowski:

<u>Theorem 3.6:</u> The number of facets of a Dirichlet parallelotope P is less or equal to $2(2^n-1)$.

Proof: The following proof of Voronoĭ (Voro1, No. 55) is based on theorem 3.5. The number of cosets of T' in T^n is 2^n. Excluding the coset T' there remain 2^n-1 cosets. If each of these cosets has exactly two shortest lattice vectors, $\vec{f}$ and $-\vec{f}$ respectively, then there are $2(2^n-1)$ facet vectors.

A d-face F of a polytope in E^n is defined by at least n-d facets (cf. section 2.4). It follows that in a face-to-face tiling of E^n at least n-d+1 contiguous polytopes meet in every d-face. Using this property Voronoĭ (Voro1, No. 16) defined the primitive Dirichlet parallelotopes.

<u>Definition 3.4:</u> A Dirichlet parallelotope P in E^n is called primitive, if in every d-face of P exactly n-d+1 contiguous Dirichlet parallelotopes meet.

The following theorem, due to Voronoĭ (Voro1, No. 17), holds:

Theorem 3.7: A Dirichlet parallelotope P is primitive if and only if in each vertex of P n+1 contiguous Dirichlet parallelotopes meet.

Proof: By assumption in each vertex v_i of P_1 exactly n+1 contiguous P_ν, $\nu=1,\ldots,n+1$, meet. Take a 1-face $E_j \supset v_i$. As v_i is a vertex of a convex polytope E_j cannot be contained in all the P_ν. Inversely E_j has to be contained in at least n of the P_ν therefore, exactly one P_ν, say P_{n+1} exists which does not contain E_j. Take a 2-face $F_k \supset E_j$. Using the same reasoning as before exactly one of the P_ν say P_n does not contain F_k. The same arguments hold for each d-face. Finally the facet F_s is exactly contained in P_1 and P_2.

It is useful to characterize facet vectors also in terms of shortest linearly independent lattice vectors. Let $|\vec{t}_1|\leq\ldots\leq|\vec{t}_n|$ be shortest linearly independent lattice vectors. They fall into k classes M_i, $i=1,\ldots,k$, of vectors of equal length. An order among the classes M_i is given by the condition that for $\vec{t}_1\in M_i$ and $\vec{t}_2\in M_{i+1}$ we assume that $|\vec{t}_1|<|\vec{t}_2|$.

Theorem 3.8: All lattice vectors of class M_1 are facet vectors. Those lattice vectors of class M_i, $i=2,\ldots,k$, which are linearly independent of the lattice vectors in classes $M_1,\ldots,M_{i-1}$ are facet vectors.

Proof: 1. We show that all lattice vectors of class M_1 are facet vectors. We take a lattice vector $\vec{f}_1\in M_1$ and assign two balls $B_0(0,|\vec{f}_1|)$ and $B_1(f_1,|\vec{f}_1|)$ of radius $|\vec{f}_1|$ and centers at the origin 0 and f_1 respectively as shown in Figure 3.5. By assumption $\mathrm{int}B_0 \cap T^n=\{0\}$. Then $\mathrm{int}B_1 \cap T^n=\{\vec{f}_1\}$ because for any other lattice point $t\in \mathrm{int}B_1$ it follows that $|\vec{t}-\vec{f}_1|<|\vec{f}_1|$ which contradicts the assumption that $\vec{f}_1\in M_1$. It follows that $\mathrm{int}(B_0\cup B_1) \cap T^n=\{0,\vec{f}_1\}$. The bisecting hyperplane H_1^o of $\vec{f}_1$ intersects $\vec{f}_1$ at $h_1:=\frac{1}{2}f_1$. We show that H_1^o generates the facet F_1. There exists a (n-1)-dimensional ball $B'(h_1,r_1) \subset H_1^o$ of radius $r_1=\sqrt{2}/2|\vec{f}_1|$ such that the bisecting hyperplane H_j^o of $\vec{t}_j$, $t_j\in \mathrm{bd}B_0 \cap H_1^o$, touches B'. By rotation symmetry around the lattice vector $\vec{f}_1$ this is true for every $t_j\in \mathrm{bd}B_0 \cap H_1^o$. The radius r_1 is minimal because for every other $t_j\notin \mathrm{int}(B_0\cup B_1)$ it follows that H_j^o avoids B'. Therefore $B'\subset F_1$ and hence, $\vec{f}_1$ is a facet vector.

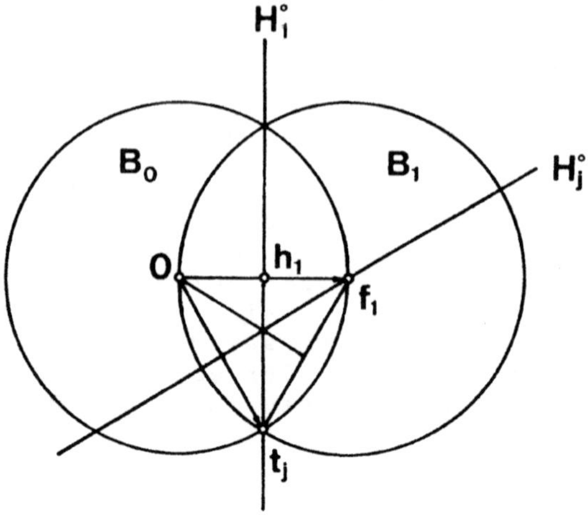

Figure 3.5

2. We now demonstrate the second part of the theorem. Let all lattice vectors $\vec{t}_j \in M_1 \cup \ldots \cup M_{i-1}$ span a sublattice T^k in E^k with lattice basis $|\vec{a}_1| \le \ldots \le |\vec{a}_k|$. We choose a lattice vector $\vec{f}_i \in M_i$ such that $\vec{f}_i \notin E^k$ and assign the balls $B_0(0,|\vec{f}_i|)$ and $B_i(f_i,|\vec{f}_i|)$. Let $M' := B_0 \cap T^k$ and let P^k be the Dirichlet parallelotope of T^k in E^k. For the region Q^k (cf. section 2.1),

$$Q^k := \bigcup_{x_\nu \in P^k} B^k(x_\nu, |\vec{x}_\nu|),$$

it holds that $\mathrm{int}Q^k \cap T^k = \{0\}$. Let p_i be the orthogonal projection of f_i into E^k. It follows that $p_i \in P^k$ because otherwise a shorter lattice vector in the coset $\vec{f}_i + T^k$ exists which contradicts the assumption that $\vec{f}_i \in M_i$. We assign the k-dimensional ball $B^k(p_i, |\vec{p}_i|) = B_i \cap E^k$. As $p_i \in P^k$ it follows that $B^k \subset Q^k$ hence, $\mathrm{int}B_i \cap T^k = \emptyset$ and therefore, $\mathrm{int}(B_0 \cup B_i) \cap T^n = \{M', \vec{f}_i\}$. The bisecting hyperplane H_i° of $\vec{f}_i$ intersects $\vec{f}_i$ at $\frac{1}{2}f_i$. We show that H_i° generates a facet F_i. There exists a (n-1)-dimensional ball $B'(\frac{1}{2}f_i, r_1)$ of radius at least r_1 such that the bisecting hyperplane H_j° of $\vec{t}_j$, $t_j \in \mathrm{bd}B_0 \cap H_i^\circ$, avoids B'. Hence, $B' \subset F_i$ and therefore, $\vec{f}_i$ is a facet vector.

Of particular importance are the zones and the belts of a parallelotope.

__Definition 3.5:__ A zone of a parallelotope P in E^n is a class of parallel 1-faces of P.

__Definition 3.6:__ A belt of a parallelotope P in E^n is a class of parallel (n-2)-faces of P.

In E^3 zones and belts are identical. The five combinatorial types of parallelohedra in E^3 were for the first time determined by Fedorov (Fedo1). These parallelohedra are shown in Figure 3.6. In each of these parallelohedra always four or six edges are parallel so they define a zone or a belt respectively.

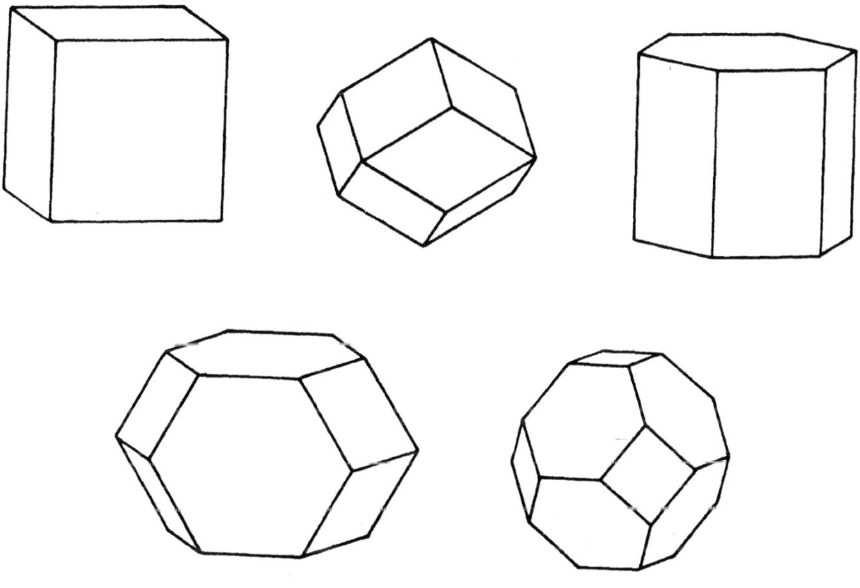

Figure 3.6. The five combinatorial types of parallelohedra in E^3

__Theorem 3.9:__ Each belt of a Dirichlet parallelotope P contains four or six (n-2)-faces.

Proof: Assume $\vec{f}_1$, $\vec{f}_2$ to be facet vectors. They generate a plane $\pi(\vec{f}_1,\vec{f}_2)$ which we call the supporting plane of the belt. If $\vec{f}_1 \bullet \vec{f}_2 < 0$ then $|\vec{f}_1+\vec{f}_2| < |\vec{f}_1-\vec{f}_2|$. It follows that

$$\tfrac{1}{2}|\vec{f}_1+\vec{f}_2| < |\vec{f}_1-\tfrac{1}{2}(\vec{f}_1+\vec{f}_2)| = \tfrac{1}{2}|\vec{f}_1-\vec{f}_2| = |\vec{f}_2-\tfrac{1}{2}(\vec{f}_1+\vec{f}_2)|.$$

Therefore, $\frac{1}{2}(f_1+f_2) \in \mathrm{int}P$ hence, $\vec{f}_1+\vec{f}_2$ is a facet vector. If $\vec{f}_1 \bullet \vec{f}_2 > 0$ then by the same argument $\vec{f}_1+\vec{f}_2$ is no facet vector. By theorem 3.4, the Dirichlet parallelotope P is centrosymmetric hence, there exist, if $\vec{f}_1 \bullet \vec{f}_2 = 0$, four or, if $\vec{f}_1 \bullet \vec{f}_2 \neq 0$, six facet vectors in the supporting plane. Let $F_1, \ldots, F_k, F_{k+1}=F_1$, $k=4$ or $k=6$, be the facets of a belt. If $F_1 \cap F_2 = F_1'$ is a (n-2)-face of P then F_1' defines a belt. By theorem 3.4, each facet is centrosymmetric therefore, also $F_i \cap F_{i+1} = F_i'$, for $i=2,..,k$. The F_i' are perpendicular to the supporting plane.

The more general result that every convex polytope which fulfils the conditions of theorem 3.4 and 3.9 is a parallelotope was proved by Venkov (Venk1) and independently by McMullen (McMu1).

The following theorem only holds for Dirichlet parallelotopes.

<u>Theorem 3.10:</u> Each belt of a Dirichlet parallelotope P is contained in a rotation cylinder which is orthogonal to the corresponding supporting plane.

Proof: This theorem immediately follows from the property that the facets are orthogonal to the facet vectors.

<u>Theorem 3.11:</u> The number of belts in a Dirichlet parallelotope in E^n, $n \geq 2$, is less or equal to

$$N = \sum_{k=2}^{n} [k\binom{}{k} + \sum_{l=2}^{n} \varepsilon \binom{n}{k}\binom{n-k}{l}] - \binom{n}{2},$$

where $\varepsilon = 1$ if $l<k$ or $\varepsilon = \frac{1}{2}$ if $l=k$ and $\binom{n-k}{l}=0$ if $n-k<l$.

The upper limit N is realized for primitive parallelotopes. The numbers N for dimension 2 to 9 are given in Table 3.1. The lower limit n is realized for the hypercube.

Proof: Let $|\vec{f}_1|\leq\ldots\leq|\vec{f}_n|$ be shortest linearly independent lattice vectors. We assume that $\vec{f}_i\cdot\vec{f}_j<0$ for $i\neq j$. By theorem 3.8, $\vec{f}_i,\vec{f}_j$ are facet vectors hence, by theorem 3.9 $\vec{f}_i+\vec{f}_j$ is also a facet vector (we note that if $\vec{f}_i\cdot\vec{f}_j>0$ then $\vec{f}_i-\vec{f}_j$ is a facet vector). It follows, in general, that also $\vec{f}_i+\vec{f}_j+\ldots+\vec{f}_k$, $i\neq j\neq\ldots\neq k$, is a facet vector.

1. The lattice vectors $\vec{f}_i,\vec{f}_j$ generate a belt. There are $\binom{n}{2}$ such belts. The lattice vectors $\vec{f}_i+\vec{f}_j$, $\vec{f}_k+\vec{f}_l$ $i,j\neq k,l$ generate a belt. There are $\binom{n}{2}$ lattice vectors $\vec{f}_i+\vec{f}_j$. To each such lattice vector exists $\binom{n-2}{2}$ different lattice vectors $\vec{f}_k+\vec{f}_l$. Thereby, we have counted every combination twice hence, there are $\frac{1}{2}\binom{n}{2}\binom{n-2}{2}$ such belts.

2. Suppose the lattice vectors $\vec{f}_i$, $\vec{f}_j+\vec{f}_k+\vec{f}_l$, $i=j$ or k or l, generate a belt. There are $\binom{n}{3}$ lattice vectors $\vec{f}_j+\vec{f}_k+\vec{f}_l$. To each such lattice vector exist three vectors $\vec{f}_i$, $i=j,k,l$. Therefore, there are $3\binom{n}{3}$ such belts. (The case that $i\neq j,k,l$ is contained in the next step where $\vec{f}_i$, $\vec{f}_j+\vec{f}_k+\vec{f}_l+\vec{f}_m$, $i=j$ or k or l or m generate a belt). Suppose the lattice vectors $\vec{f}_i+\vec{f}_j$, $\vec{f}_k+\vec{f}_l+\vec{f}_m$, $i,j\neq k,l,m$ generate a belt. There are $\binom{n}{3}$ lattice vectors $\vec{f}_k+\vec{f}_l+\vec{f}_m$. To each such lattice vector exist $\binom{n-3}{2}$ lattice vectors $\vec{f}_i+\vec{f}_j$. Hence, there are at most $\binom{n}{3}\binom{n-3}{2}$ such belts. Suppose the lattice vectors $\vec{f}_i+\vec{f}_j+\vec{f}_k$, $\vec{f}_l+\vec{f}_m+\vec{f}_n$, $i,j,k\neq l,m,n$ generate a belt. There are $\binom{n}{3}$ lattice vectors $\vec{f}_i+\vec{f}_j+\vec{f}_k$. To each such lattic vector exists $\binom{n-3}{3}$ vectors $\vec{f}_l+\vec{f}_m+\vec{f}_n$. Thereby, we have counted every combination twice hence, there are $\frac{1}{2}\binom{n}{3}\binom{n-3}{3}$ such belts.

3. Finally suppose $\vec{f}_i$, $\vec{f}_1+\ldots+\vec{f}_n$ generate a belt. There is $\binom{n}{n}=1$ vector $\vec{f}_1+\ldots+\vec{f}_n$. There exist n lattice vectors $\vec{f}_i$, $i=1,\ldots,n$. Hence, there are n such belts.

Table 3.1. The maximal number N of belts of a Dirichlet parallelotope in E^n

n	2	3	4	5	6	7	8	9
N	1	6	25	90	301	966	3025	9330

The next theorem reveals the important property that always a lattice basis can be found among the facet vectors of a Dirichlet parallelotope $P(x_0)$. For a vertex $v_h\in clP(x_0)$ we assign the ball $B(v_h,|\vec{v}_h|)$ of radius $|\vec{v}_h|$.

The bisecting hyperplanes H_i of all lattice vectors $\vec{f}_i \in \mathrm{cl}B(v_h, |\vec{v}_h|)$, $i=1,\ldots,l$, contain the vertex v_h. Let k of these lattice vectors, $\vec{f}_1,\ldots,\vec{f}_k,\vec{f}_{k+1}=\vec{f}_1$, be facet vectors. We cyclically order these facet vectors by the requirement that for the corresponding facets we obtain

$$F_i \cap F_{i+1} = F'_i \subset \mathrm{cl}P(x_0), \quad i=1,\ldots,k,$$

where F'_i is a (n-2)-face of $\mathrm{cl}P(x_0)$. Such an ordered set of facet vectors we call a cycle of facet vectors belonging to a vertex v. We generate the orbit of v_h under the group G of all isometries of the lattice T^n (cf. section 8.1),

$$O(G,v_h) := \{v_\nu \mid v_\nu = \sigma_\nu v_h, \ \sigma_\nu \in G\}.$$

Let $D(v_h)$ be the Dirichlet domain of v_h with respect to $O(G,v_h)$. Let R_m be the radius of a maximal ball of $D(v_h)$.

<u>Theorem 3.12:</u> If $|\vec{v}_h|>R_m$ then in the cycle of facet vectors $\vec{f}_1,\ldots,\vec{f}_k,\vec{f}_{k+1}=\vec{f}_1$, belonging to the vertex $v_h \in \mathrm{cl}P(x_0)$ any sequence of n consecutive facet vectors forms a lattice basis of T^n.

Proof: 1. First we show that the lattice T' generated by $\vec{f}_1,\ldots,\vec{f}_l \in \mathrm{cl}B(v_h,|\vec{v}_h|)$ is not a sublattice of T^n. Suppose that $T' \subset T^n$. The coset decomposition of T^n becomes

$$T^n = T' \cup \vec{b}_1 + T' \cup \ldots \cup \vec{b}_m + T'.$$

We attach at each vertex $v_\nu \in O(G,v_h)$ a ball $B(v_\nu,|\vec{v}_h|)$. By the covering property (cf. theorem 2.5) it follows that

$$\bigcup_{v_\nu \in O(G,v_h)} \mathrm{cl}B(v_\nu,R_m) = E^n.$$

By the condition $|\vec{v}_h|>R_m$ it follows that

$$\bigcup_{v_\nu \in O(G,v_h)} B(v_\nu,|\vec{v}_h|) = E^n.$$

Therefore, a ν exists such that the coset representative $\vec{b}_i$ is contained in $B(v_\nu,|\vec{v}_h|)$. Since v_ν is equivalent to v_h under G there exists an isometry $\sigma \in G$ which carries $\vec{b}_i$ into $\vec{b}' \in B(v_h,|\vec{v}_h|)$. Therefore, the bisecting hyperplane H^o of $\vec{b}'$ cuts off v_h hence, $v_h \notin \mathrm{cl}P(x_0)$ which contradicts the assumption that $v_h \in clP(x_0)$. It follows that $T'=T^n$.

2. Next we show that any sequence of n consecutive facet vectors among the cycle $\vec{f}_1,\ldots,\vec{f}_k,\vec{f}_{k+1}=\vec{f}_1$ form a lattice basis. Always n among them are linearly independent because otherwise v_h cannot be a vertex of a convex polytope. We choose any sequence of n facet vectors which we designate with $\vec{a}_1,\ldots,\vec{a}_n$. The corresponding facets are called $F_1,\ldots,F_n$. We take $\vec{a}_1$. Translates of $clP(x_0)$ by $\vec{a}_1$ form a prism $C(\vec{a}_1)$ in E^n which is topologically equivalent to a cylinder $C' \subset E^n$,

$$C' := \lambda_1\vec{a}_1 + \sum_{j=2}^{n} \lambda_j\vec{a}_j,\ \sum_{j=2}^{n} \lambda_j^2=1.$$

λ_1 is a free parameter. In general, the number of free paramters λ_i is defined as the extent of C', abbreviated as $ext(C')$. In this case $ext(C')=1$. By assumption $F_1 \cap F_2=F_1'$. The (n-2)-face F_1' defines a belt. By theorem 3.8, the projection of the belt into the supporting plane $\pi(\vec{a}_1,\vec{a}_2)$ is either a centrosymmetric hexagon or rectangle. Both are parallelogons. Translates of these with $\vec{a}_1$ and $\vec{a}_2$ cover the plane $\pi(\vec{a}_1,\vec{a}_2)$. The corresponding Dirichlet parallelotopes form a prism $C(\vec{a}_1,\vec{a}_2)$ of extent 2. We take $\vec{a}_3$. By assumption $F_2 \cap F_3=F_2'$. The (n-2)-face F_2' defines a belt. The parallelogons in the supporting plane $\pi(\vec{a}_2,\vec{a}_3)$ cover the plane. The corresponding Dirichlet parallelotopes form a prism $C(\vec{a}_2,\vec{a}_3)$. By multiples of the translation $\vec{a}_1$ this is true for every $P \in C(\vec{a}_1)$. Thus we get a prism $C(\vec{a}_1,\vec{a}_2,\vec{a}_3)$ of extent 3. This process is continued until $\vec{a}_n$. Finally we have $C(\vec{a}_1,\ldots,\vec{a}_n)=E^n$. Thus $\vec{a}_1,\ldots,\vec{a}_n$ form a lattice basis.

Let R be the radius of a maximal ball of the Dirichlet parallelotope $P(x_0)$, $R=\max(|\vec{v}_i|,\ v_i \in clP(x_0))$. As $D(v_h) \subset P(v_h)$, $P(v_h):=P(x_0)+\vec{v}_h$, it follows that $R_m \leq R$ therefore, for every lattice T^n there exists a lattice basis of facet vectors. For $R_m=R$ the equal sign holds in theorem 3.12.

4. Reduction of Quadratic Forms

For applications it is important to decide wether two different lattice bases belong to one and the same lattice T^n. This problem is solved through the determination of an integral reduced form denoted here as $\mathbb{Z}$-reduced.

4.1. Definition of the $\mathbb{Z}$-reduced form

In section 3.2 we have defined the positive-definite n-nary quadratic form

$$\vec{x}^t C \vec{x} = \sum_{i=1}^{n} \sum_{j=1}^{n} c_{ij} m_i m_j = f(m_1, \ldots, m_n),$$

where C is the metric tensor of a lattice basis of T^n. Under a change of lattice basis the metric tensor C is transformed according to (cf. section 3.2)

$$C' := ACA^t,$$

where A is a unimodular nxn matrix with integral coefficients. This is an equivalence relation and C' is said to be equivalent to C (cf. section 1.5).

Definition 4.1: In a class of equivalent positive-definite n-nary quadratic forms a form is called $\mathbb{Z}$-reduced if it assumes successive minima for the n-tuples (1,0,...,0),... ,(0,...,0,1) and, if some of the minima are not unique, the form fulfils a system of selection rules.

The following theorem is due to Lagrange (for n=2), to Seeber (for n=3) and, in the general case, to Minkowski (Mink1).

Theorem 4.1: In each class of equivalent positive-definite n-nary quadratic forms exists a unique $\mathbb{Z}$-reduced form.

Proof: Each class of equivalent positive-definite n-nary quadratic forms represents a lattice T^n. Let P_0 be the parallelotope of T^n and let R be the radius of a maximal interstitial ball of T^n. For n=1 the theorem is evident. Take n>1. We assume the theorem to be correct for $n-1$. Let P' be the parallelotope of T^{n-1} and let $R'<R$ be the radius of a maximal interstitial ball of T^{n-1}. Within a layer spacing of $2R$ from T^{n-1} at least another coset $U:=\vec{t}+T^{n-1}$ exists. There exists a lattice vector $\vec{a}_n \in U$ such that the orthogonal projection of $\vec{a}_n$ falls into the interior or on the boundary of P'. It follows that $|\vec{a}_n| \le \sqrt{4R^2+R'^2} < \sqrt{5}R$. Within a ball of radius $\sqrt{5}R$ only a finite number of lattice vectors exists therefore, a lattice vector $\vec{a}_n$ of minimal length can be found. If $\vec{a}_n$ is not unique then a finite number of selection rules can specify a unique solution.

In the following sections we will discuss various reduction schemes. There are other possibilities to define a $\mathbb{Z}$-reduced form. Such an alternative definition is given in section 4.4.

4.2. The reduction scheme of Lagrange

The reduction scheme of Lagrange is based on the follwo-ing theorem:

Theorem 4.2: In a lattice T^2 two shortest linearly independent lattice vectors form a lattice basis.

Proof: Let $|\vec{a}_1| \le |\vec{a}_2|$ be two shortest linearly independent lattice vectors. Suppose they generate a lattice T'^2. We construct the Dirichlet parallelogon P_0 of T'^2. Let v_h be a vertex of $\mathrm{cl}P_0$. By theorem 3.10, all vertices of $\mathrm{cl}P_0$ lie on a circle hence, $R=|v_h|$ is the radius of the maximal interstitial ball. If $\vec{a}_1, \vec{a}_2$ do not form a lattice basis of T^2 then necessarily there exists a further lattice vector $\vec{t} \in \mathrm{cl}P_0$. By theorem 2.4, it follows that

$$|\vec{t}| \le R \le \tfrac{1}{2}\sqrt{|\vec{a}_1|^2+|\vec{a}_2|^2} \le \sqrt{2}/2\,|\vec{a}_2|.$$

Hence, $|\vec{t}|<|\vec{a}_2|$ which is a contradiction.

For the derivation of Lagrange's reduction conditions we can assume that

$$0 < c_{11} \le c_{22}.$$

Then necessarily, as can be seen from Figure 4.1, it holds that

$$2|c_{12}| \le c_{11}$$

because otherwise

$$\vec{a}_2' := \vec{a}_2 - \vec{a}_1 \operatorname{sgn}(c_{12})$$

would be shorter than $\vec{a}_2$ since

$$c_{22}' := c_{11} + c_{22} - 2|c_{12}|$$

by the law of cosines. The expression $\operatorname{sgn}(c_{12})$ denotes the sign of c_{12} and may assume the values +1 if $c_{12}>0$, 0 if $c_{12}=0$, or −1 if $c_{12}<0$. If further we require that the angle α_{12} is acute then according to Lagrange (Lagr1) a complete system of reduction conditions in E^2 is obtained.

Definition 4.2: A positive-definite binary quadratic form is called $\mathbb{Z}$-reduced according to Lagrange if the following conditions are fulfilled:

$$0 \le 2c_{12} \le c_{11} \le c_{22}$$

4.3. The reduction scheme of Seeber

4.3.1. Reduction conditions

The reduction scheme of Seeber is based on the following theorem:

Theorem 4.3: In a lattice T^3 three shortest linearly independent lattice vectors form a lattice basis.

Proof: Let $|\vec{a}_1|\le|\vec{a}_2|\le|\vec{a}_3|$ be three shortest linearly independent lattice vectors. Suppose they generate a lattice T'^3. Let P_0 be the Dirichlet parallelohedron of T'^3 and let R be the radius of a maximal interstitial ball. If $\vec{a}_1$, $\vec{a}_2$, and $\vec{a}_3$ do not form a lattice basis of T^3 then necessarily there exists a further lattice vector $\vec{t}\in \mathrm{cl}P_0$. By theorem 2.4, it follows that

$$|\vec{t}|\le R\le\tfrac{1}{2}\sqrt{|\vec{a}_1|^2+|\vec{a}_2|^2+|\vec{a}_3|^2} \le \sqrt{3}/2|\vec{a}_3|.$$

Hence, $|\vec{t}|<|\vec{a}_3|$ which is a contradiction.

If some of the successive minima of the form are not unique then a rather complicated set of specification rules is needed in order to obtain a complete system of reduction conditions. The following reduction scheme in E^3 is due to Seeber (Seeb2, No. 26). The angles between the basis vectors can be chosen either all acute or else all obtuse or right. Correspondingly a $\mathbb{Z}$-reduced form is called here acute or obtuse respectively (Seeber called it "positive" or "negative" respectively which however is confusing with the expression positive-definite).

Definition 4.3: A positive-definite ternary quadratic form is called $\mathbb{Z}$-reduced according to Seeber (also Niggli-reduced) if the following conditions are fulfilled:

1. For acute $\mathbb{Z}$-reduced forms with $c_{12}, c_{13}, c_{23} > 0$
   Main conditions:
   $c_{11} \le c_{22} \le c_{33}$
   $2c_{12} \le c_{11},\ 2c_{13} \le c_{11},\ 2c_{23} \le c_{22}$

   Auxillary conditions:
   $c_{23} \le c_{13}$ if $c_{11} = c_{22}$
   $c_{13} \le c_{12}$ if $c_{22} = c_{33}$
   $c_{12} \le 2c_{13}$ if $2c_{23} = c_{22}$
   $c_{12} \le 2c_{23}$ if $2c_{13} = c_{11}$
   $c_{13} \le 2c_{23}$ if $2c_{12} = c_{11}$

2. For obtuse $\mathbb{Z}$-reduced forms with $c_{12}, c_{13}, c_{23} \le 0$
   Main conditions:
   $c_{11} \le c_{22} \le c_{33}$
   $2|c_{12}| \le c_{11},\ 2|c_{13}| \le c_{11},\ 2|c_{23}| \le c_{22}$
   $2(|c_{12} + c_{13} + c_{23}|) \le c_{11} + c_{22}$

   Auxillary conditions:
   $|c_{23}| \le |c_{13}|$ if $c_{11} = c_{22}$
   $|c_{13}| \le |c_{12}|$ if $c_{22} = c_{33}$
   $|c_{12}| = 0$ if $2|c_{23}| = c_{22}$
   $|c_{12}| = 0$ if $2|c_{13}| = c_{11}$
   $|c_{13}| = 0$ if $2|c_{12}| = c_{11}$
   $c_{11} \le |c_{12}| + 2|c_{13}|$ if $2|c_{12} + c_{13} + c_{23}| = c_{11} + c_{22}$

This system of reduction conditions determines for each class of equivalent forms a unique $\mathbb{Z}$-reduced lattice basis.

A geometric interpretation of Seeber's reduction conditions was given by Dirichlet (Diri1) who introduced the

Dirichlet domain. By theorem 4.2, two shortest linearly independent lattice vectors $\vec{a}_1$, $\vec{a}_2$ generate a sublattice T^2 with Dirichlet parallelogon P. The orthogonal projection of a shortest lattice vector $\vec{a}_3$, which is linearly independent of $\vec{a}_1$ and $\vec{a}_2$, has to fall into the interior or on the boundary of P because otherwise a shorter lattice vector would exist in the coset $\vec{a}_3+T^2$. It follows that the boundary of P gives the required conditions for a shortest lattice vector $\vec{a}_3$. If $\vec{t}$ generates an edge of P then necessarily

$$|\vec{a}_3\bullet\vec{t}|\leq\tfrac{1}{2}|\vec{t}|^2.$$

By theorem 3.9, the parallelogon P has four or six edges only. It follows that for acute reduced forms with $\vec{a}_1\bullet\vec{a}_2>0$ only the two conditions

$$|\vec{a}_1\bullet\vec{a}_3|\leq\tfrac{1}{2}|\vec{a}_1|^2 \quad\text{and}\quad |\vec{a}_2\bullet\vec{a}_3|\leq\tfrac{1}{2}|\vec{a}_2|^2$$

have to be considered for $\vec{a}_3$. In terms of the metric tensor these conditions are written as

$$2|c_{13}|\leq c_{11} \quad\text{and}\quad 2|c_{23}|\leq c_{22}.$$

For obtuse reduced forms with $\vec{a}_1\bullet\vec{a}_2<0$ also the edge determined by the vector $\vec{a}_1+\vec{a}_2$ gives a possible condition for $\vec{a}_3$ hence, in addition we have

$$|\vec{a}_1\bullet\vec{a}_3+\vec{a}_2\bullet\vec{a}_3|\leq\tfrac{1}{2}|\vec{a}_1+\vec{a}_2|^2=\tfrac{1}{2}(|\vec{a}_1|^2-2|\vec{a}_1\bullet\vec{a}_2|+|\vec{a}_2|^2).$$

In terms of the metric tensor this condition can be written as

$$2|c_{12}+c_{13}+c_{23}|\leq c_{11}+c_{22}.$$

If the orthogonal projection of $\vec{a}_3$ falls on the boundary of P then auxillary conditions are needed in order to make a unique choice (for details the reader is referred to the International Tables Vol. A).

4.3.2. Determination of the Seeber reduced form

Seeber described the following algorithm to determine the reduced lattice basis (Seeb2, No. 27).

If $2|c_{12}| > c_{11}$ then a reduction step in Seeber's algorithm is as follows: We set

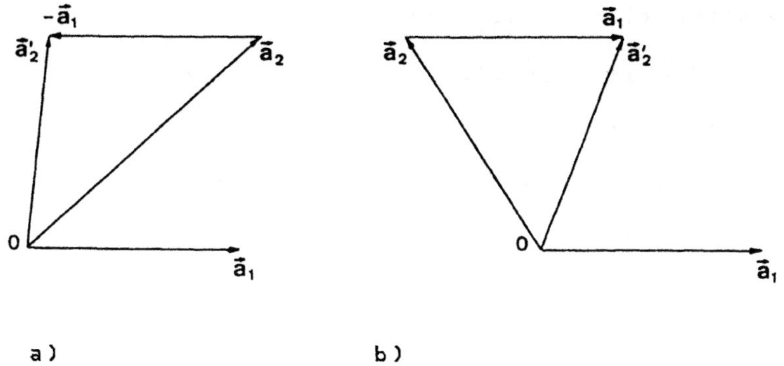

Figure 4.1. A reduction step in Seeber's algorithm
a) $c_{12}>0$, b) $c_{12}<0$

$$\vec{a}'_2 := \vec{a}_2 - \vec{a}_1 \operatorname{sgn}(c_{12}).$$

Then by the law of cosines

$$|\vec{a}'_2|^2 = |\vec{a}_1|^2 + |\vec{a}_2|^2 - 2|\vec{a}_1|\ |\vec{a}_2|\ \cos\alpha_{12}\ \operatorname{sgn}(c_{12}).$$

It follows that

$$c'_{22} := c_{11} + c_{22} - 2|c_{12}| < c_{22}.$$

The new angle α'_{2i} between the lattice vectors $\vec{a}'_2$ and $\vec{a}_i$ is obtained from

$$c'_{2i} := \vec{a}_i \bullet \vec{a}'_2 = \vec{a}_i \bullet (\vec{a}_2 - \vec{a}_1 \operatorname{sgn}(c_{12}))$$

$$= c_{2i} - c_{1i}\ \operatorname{sgn}(c_{12}),\ i=1,3.$$

In matrix notation this transformation writes $C' = ACA^t$, with

$$A := \begin{vmatrix} 1 & 0 & 0 \\ -\operatorname{sgn}(c_{12}) & 1 & 0 \\ 0 & 0 & 1 \end{vmatrix} .$$

Similarly, if $2|c_{13}| > c_{11}$ we have the transformation

$$A := \begin{vmatrix} 1 & 0 & 0 \\ 0 & 1 & 0 \\ -\operatorname{sgn}(c_{13}) & 0 & 1 \end{vmatrix} .$$

This transformation results in

$$c'_{33} := c_{11} + c_{33} - 2|c_{13}|$$

$$c'_{3i} := c_{3i} - c_{1i}\ \mathrm{sgn}(c_{13}),\ i=1,2.$$

If $2|c_{23}| > c_{22}$ we have the transformation

$$A := \begin{vmatrix} 1 & 0 & 0 \\ 0 & 1 & 0 \\ 0 & -\mathrm{sgn}(c_{23}) & 1 \end{vmatrix}$$

which results in

$$c'_{33} := c_{22} + c_{33} - 2|c_{23}|$$
$$c'_{3i} := c_{3i} - c_{2i}\ \mathrm{sgn}(c_{23}),\ i=1,2.$$

An analogous algorithm was described independently by Křivý and Gruber in 1976 (Křiv1).

Using Seeber's reduction conditions Niggli (Nigg3) investigated the lattices in E^3 and distinguished 44 types of Seeber $\mathbb{Z}$-reduced forms which are given in Table 4.2.

For the reduction of positive-definite ternary quadratic forms a computer program has been developed (Enge6). The program gives after m reduction steps the final transformation

$$A_m \ldots A_1,$$

where the A_i are the transformations given in Seeber's algorithm. The program then assigns the $\mathbb{Z}$-reduced form to one of Niggli's 44 types and gives the transformation W into the conventional crystallographic unit cell according to the International Tables Vol. A (Hahn1).

Table 4.1. The Křivý–Gruber algorithm to determine the Seeber $\mathbb{Z}$-reduced positive ternary quadratic form

1. If $c_{11} > c_{22}$ or ($c_{11} = c_{22}$ and $|c_{23}| > |c_{13}|$) then
   exchange (c_{11}, c_{23}) with (c_{22}, c_{13}).

2. If $c_{22} > c_{33}$ or ($c_{22} = c_{33}$ and $|c_{13}| > |c_{12}|$) then
   exchange (c_{22}, c_{13}) with (c_{33}, c_{12}).
   Go to 1.

3. If $c_{12} * c_{13} * c_{23} > 0$
   then set $c_{12} := |c_{12}|$, $c_{13} := |c_{13}|$, $c_{23} := |c_{23}|$.
   else set $c_{12} := -|c_{12}|$, $c_{13} := -|c_{13}|$, $c_{23} := -|c_{23}|$.

4. If $2|c_{23}| > c_{22}$ or ($2c_{23} = c_{22}$ and $2c_{13} < c_{12}$) or
   ($2c_{23} = -c_{22}$ and $c_{12} < 0$)
   then set $c_{33} := c_{22} + c_{33} - 2|c_{23}|$,
   $c_{13} := c_{13} - c_{12} * \mathrm{sgn}(c_{23})$, $c_{23} := c_{23} - c_{22} * \mathrm{sgn}(c_{23})$.
   Go to 1.

5. If $2|c_{13}| > c_{11}$ or ($2c_{13} = c_{11}$ and $2c_{23} < c_{12}$) or
   ($2c_{13} = -c_{11}$ and $c_{12} < 0$)
   then set $c_{33} := c_{11} + c_{33} - 2|c_{13}|$,
   $c_{23} := c_{23} - c_{12} * \mathrm{sgn}(c_{13})$, $c_{13} := c_{13} - c_{11} * \mathrm{sgn}(c_{13})$.
   Go to 1.

6. If $2|c_{12}| > c_{11}$ or ($2c_{12} = c_{11}$ and $2c_{23} < c_{13}$) or
   ($2c_{12} = -c_{11}$ and $c_{13} < 0$)
   then set $c_{22} := c_{11} + c_{22} - 2|c_{12}|$,
   $c_{23} := c_{23} - c_{13} * \mathrm{sgn}(c_{12})$, $c_{12} := c_{12} - c_{11} * \mathrm{sgn}(c_{12})$.
   Go to 1.

7. If $2c_{12} + 2c_{13} + 2c_{23} + c_{11} + c_{22} < 0$ or
   ($2c_{12} + 2c_{13} + 2c_{23} + c_{11} + c_{22} = 0$ and $c_{11} + 2c_{13} + c_{12} > 0$)
   then set $c_{33} := c_{11} + c_{22} + c_{33} + 2c_{12} + 2c_{13} + 2c_{23}$,
   $c_{23} := c_{22} + c_{12} + c_{23}$, $c_{13} := c_{11} + c_{12} + c_{13}$.
   Go to 1.

Table 4.2. The 44 types of $\mathbb{Z}$-reduced forms according to Niggli

Bravais-lattice — $\mathbb{Z}$-reduced form (c_{11} c_{22} c_{33} / c_{12} c_{13} c_{23}) — transformation W

1. Cubic crystal system

Bravais-lattice		$\mathbb{Z}$-reduced form			transformation W
P	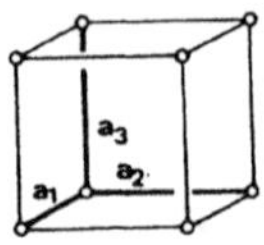	c_{11}	c_{11}	c_{11}	$\begin{vmatrix} 1 & 0 & 0 \\ 0 & 1 & 0 \\ 0 & 0 & 1 \end{vmatrix}$
		0	0	0	
I	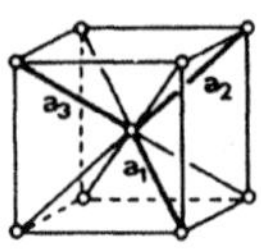	c_{11}	c_{11}	c_{11}	$\begin{vmatrix} 1 & 0 & 1 \\ 1 & 1 & 0 \\ 0 & 1 & 1 \end{vmatrix}$
		$-\frac{1}{3}c_{11}$	$-\frac{1}{3}c_{11}$	$-\frac{1}{3}c_{11}$	
F	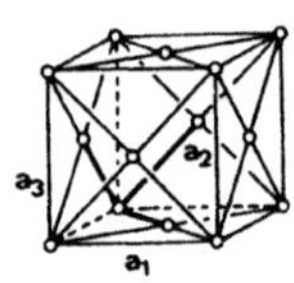	c_{11}	c_{11}	c_{11}	$\begin{vmatrix} 1 & -1 & 1 \\ 1 & 1 & -1 \\ -1 & 1 & 1 \end{vmatrix}$
		$\frac{1}{2}c_{11}$	$\frac{1}{2}c_{11}$	$\frac{1}{2}c_{11}$	

2. Hexagonal crystal system

Bravais-lattice		$\mathbb{Z}$-reduced form			transformation W
P	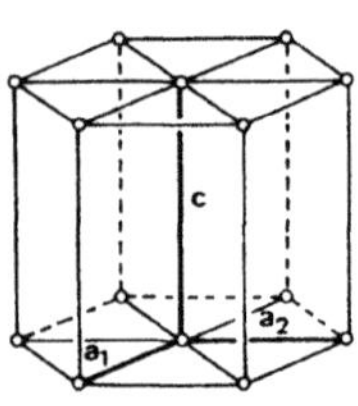	c_{11}	c_{11}	c_{33}	$\begin{vmatrix} 1 & 0 & 0 \\ 0 & 1 & 0 \\ 0 & 0 & 1 \end{vmatrix}$
		$-\frac{1}{2}c_{11}$	0	0	
P	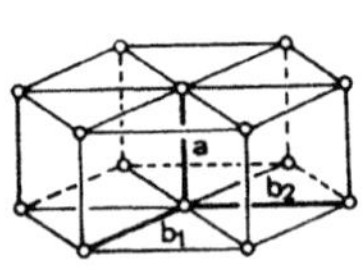	c_{11}	c_{22}	c_{22}	$\begin{vmatrix} 0 & 1 & 0 \\ 0 & 0 & 1 \\ 1 & 0 & 0 \end{vmatrix}$
		0	0	$-\frac{1}{2}c_{22}$	

Table 4.2. (continued)

Bravais-lattice | $\mathbb{Z}$-reduced form | transformation W

3. <u>Rhombohedral crystal system</u>

R

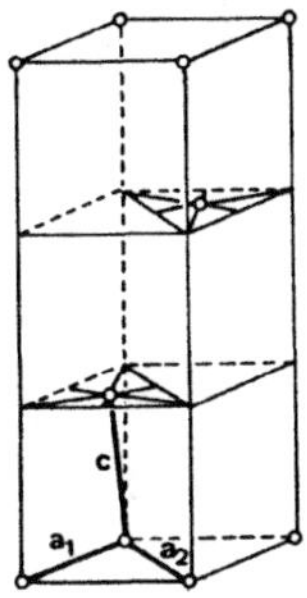

$$\begin{matrix} c_{11} & c_{11} & c_{33} \\ \frac{1}{2}c_{11} & \frac{1}{2}c_{11} & \frac{1}{2}c_{11} \end{matrix} \qquad \begin{vmatrix} 1 & 0 & 0 \\ -1 & 1 & 0 \\ -1 & -1 & 3 \end{vmatrix}$$

R

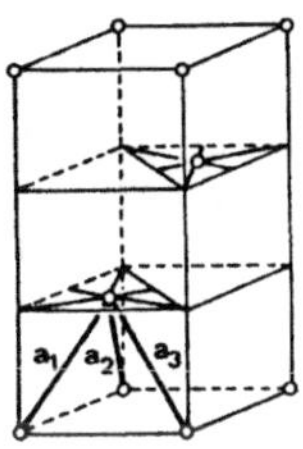

$$\begin{matrix} c_{11} & c_{11} & c_{11} \\ |c_{12}| & |c_{12}| & |c_{12}| \end{matrix} \qquad \begin{vmatrix} 1 & -1 & 0 \\ -1 & 0 & 1 \\ -1 & -1 & -1 \end{vmatrix}$$

R

$$\begin{matrix} c_{11} & c_{11} & c_{11} \\ -|c_{12}| & -|c_{12}| & -|c_{12}| \end{matrix} \qquad \begin{vmatrix} 1 & -1 & 0 \\ -1 & 0 & 1 \\ -1 & -1 & -1 \end{vmatrix}$$

R

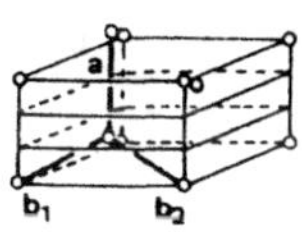

$$\begin{matrix} c_{11} & c_{22} & c_{22} \\ -\frac{1}{3}c_{11} & -\frac{1}{3}c_{11} & -\frac{1}{2}c_{22}+\frac{1}{6}c_{11} \end{matrix} \qquad \begin{vmatrix} 1 & 2 & 1 \\ 0 & -1 & 1 \\ 1 & 0 & 0 \end{vmatrix}$$

4. <u>Tetragonal crystal system</u>

P

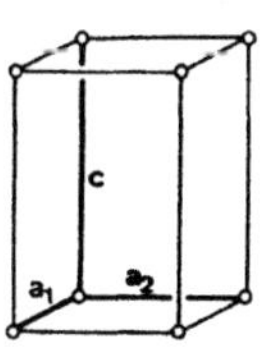

$$\begin{matrix} c_{11} & c_{11} & c_{33} \\ 0 & 0 & 0 \end{matrix} \qquad \begin{vmatrix} 1 & 0 & 0 \\ 0 & 1 & 0 \\ 0 & 0 & 1 \end{vmatrix}$$

P

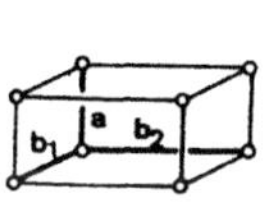

$$\begin{matrix} c_{11} & c_{22} & c_{22} \\ 0 & 0 & 0 \end{matrix} \qquad \begin{vmatrix} 0 & 1 & 0 \\ 0 & 0 & 1 \\ 1 & 0 & 0 \end{vmatrix}$$

Table 4.2. (continued)

Bravais-lattice		ℤ-reduced form	transformation W
I	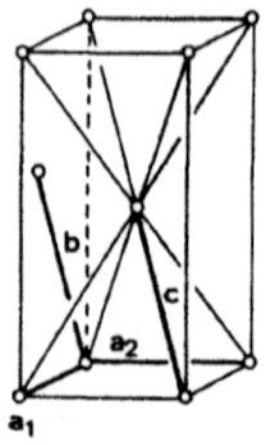	$\begin{matrix} c_{11} & c_{11} & c_{33} \\ 0 & -\frac{1}{2}c_{11} & -\frac{1}{2}c_{11} \end{matrix}$	$\begin{vmatrix} 1 & 0 & 0 \\ 0 & 1 & 0 \\ 1 & 1 & 2 \end{vmatrix}$
I	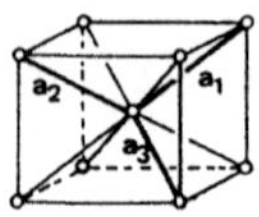	$\begin{matrix} c_{11} & c_{11} & c_{11} \\ -\lvert c_{12}\rvert & \frac{1}{2}(\lvert c_{12}\rvert-c_{11}) & \frac{1}{2}(\lvert c_{12}\rvert-c_{11}) \end{matrix}$	$\begin{vmatrix} 0 & 1 & 1 \\ 1 & 0 & 1 \\ 1 & 1 & 0 \end{vmatrix}$
I	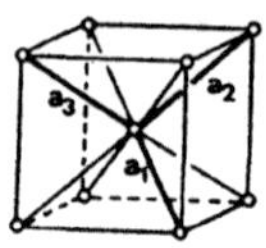	$\begin{matrix} c_{11} & c_{11} & c_{11} \\ \frac{1}{2}(\lvert c_{23}\rvert-c_{11}) & \frac{1}{2}(\lvert c_{23}\rvert-c_{11}) & -\lvert c_{23}\rvert \end{matrix}$	$\begin{vmatrix} 1 & 0 & 1 \\ 1 & 1 & 0 \\ 0 & 1 & 1 \end{vmatrix}$
I	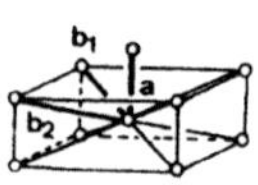	$\begin{matrix} c_{11} & c_{22} & c_{22} \\ \frac{1}{2}c_{11} & \frac{1}{2}c_{11} & \frac{1}{4}c_{11} \end{matrix}$	$\begin{vmatrix} 0 & -1 & 1 \\ 1 & -1 & -1 \\ 1 & 0 & 0 \end{vmatrix}$

5. <u>Orthorhombic crystal system</u>

Bravais-lattice		ℤ-reduced form	transformation W
P	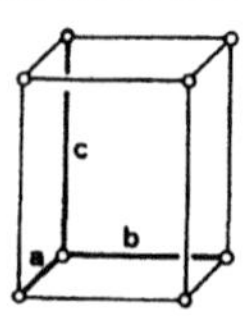	$\begin{matrix} c_{11} & c_{22} & c_{33} \\ 0 & 0 & 0 \end{matrix}$	$\begin{vmatrix} 1 & 0 & 0 \\ 0 & 1 & 0 \\ 0 & 0 & 1 \end{vmatrix}$
C	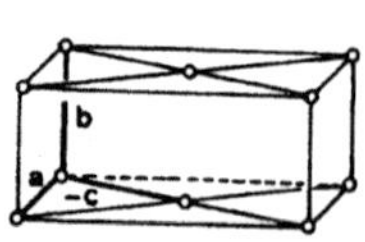	$\begin{matrix} c_{11} & c_{22} & c_{33} \\ 0 & -\frac{1}{2}c_{11} & 0 \end{matrix}$	$\begin{vmatrix} 1 & 0 & 0 \\ -1 & 0 & -2 \\ 0 & 1 & 0 \end{vmatrix}$

Table 4.2. (continued)

Bravais-lattice		$\mathbb{Z}$-reduced form	transformation W
C	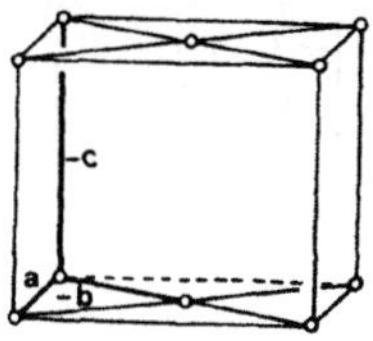	$\begin{matrix} c_{11} & c_{22} & c_{33} \\ -\frac{1}{2}c_{11} & 0 & 0 \end{matrix}$	$\begin{pmatrix} 1 & 0 & 0 \\ -1 & -2 & 0 \\ 0 & 0 & 1 \end{pmatrix}$
C	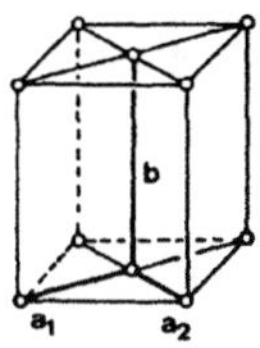	$\begin{matrix} c_{11} & c_{11} & c_{33} \\ -\lvert c_{12}\rvert & 0 & 0 \end{matrix}$	$\begin{pmatrix} 1 & 1 & 0 \\ -1 & 1 & 0 \\ 0 & 0 & 1 \end{pmatrix}$
C	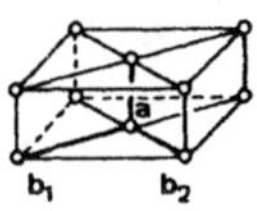	$\begin{matrix} c_{11} & c_{22} & c_{22} \\ 0 & 0 & -\lvert c_{23}\rvert \end{matrix}$	$\begin{pmatrix} 0 & 1 & 1 \\ 0 & -1 & 1 \\ 1 & 0 & 0 \end{pmatrix}$
C	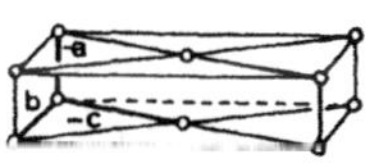	$\begin{matrix} c_{11} & c_{22} & c_{33} \\ 0 & 0 & -\frac{1}{2}c_{22} \end{matrix}$	$\begin{pmatrix} 0 & 1 & 0 \\ 0 & -1 & -2 \\ -1 & 0 & 0 \end{pmatrix}$
I	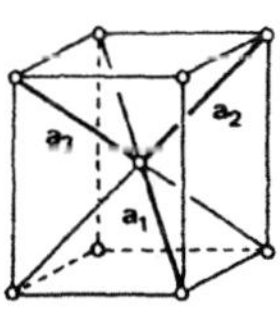	$\begin{matrix} c_{11} & c_{11} & c_{11} \\ \lvert c_{23}\rvert+\lvert c_{13}\rvert-c_{11} & -\lvert c_{13}\rvert & -\lvert c_{23}\rvert \end{matrix}$	$\begin{pmatrix} -1 & -1 & 0 \\ 1 & 0 & 1 \\ 0 & 1 & 1 \end{pmatrix}$
I	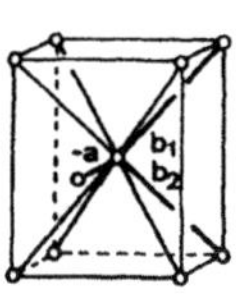	$\begin{matrix} c_{11} & c_{22} & c_{22} \\ \frac{1}{2}c_{11} & \frac{1}{2}c_{11} & \lvert c_{23}\rvert \end{matrix}$	$\begin{pmatrix} 1 & 0 & 0 \\ 0 & 1 & -1 \\ -1 & 1 & 1 \end{pmatrix}$

Table 4.2. (continued)

Bravais-lattice		$\mathbb{Z}$-reduced form	transformation W
I	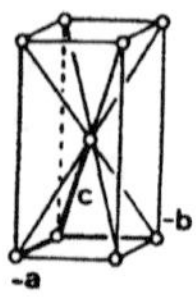	$c_{11} \quad c_{22} \quad c_{33}$ $0 \quad -\frac{1}{2}c_{11} \quad -\frac{1}{2}c_{22}$	$\begin{vmatrix} -1 & 0 & 0 \\ 0 & -1 & 0 \\ 1 & 1 & 2 \end{vmatrix}$
F	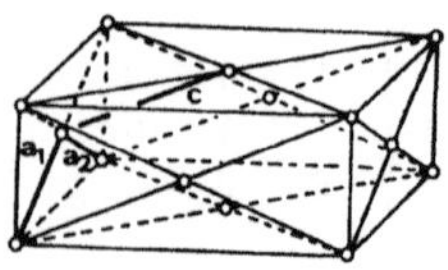	$c_{11} \quad c_{11} \quad c_{33}$ $2\lvert c_{13}\rvert - c_{11} \quad -\lvert c_{13}\rvert \quad -\lvert c_{13}\rvert$	$\begin{vmatrix} -1 & -1 & 0 \\ 1 & -1 & 0 \\ 1 & 1 & 2 \end{vmatrix}$
F	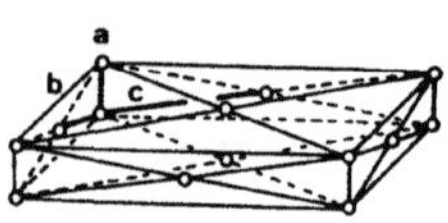	$c_{11} \quad c_{22} \quad c_{33}$ $\frac{1}{2}c_{11} \quad \frac{1}{2}c_{11} \quad \frac{1}{4}c_{11}$	$\begin{vmatrix} -1 & 2 & 0 \\ -1 & 0 & 2 \\ 1 & 0 & 0 \end{vmatrix}$

6. <u>Monoclinic crystal system</u>

Bravais-lattice		$\mathbb{Z}$-reduced form	transformation W
P	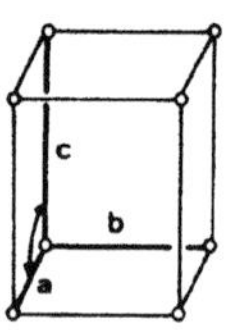	$c_{11} \quad c_{22} \quad c_{33}$ $0 \quad -\lvert c_{13}\rvert \quad 0$	$\begin{vmatrix} 1 & 0 & 0 \\ 0 & 1 & 0 \\ 0 & 0 & 1 \end{vmatrix}$
P	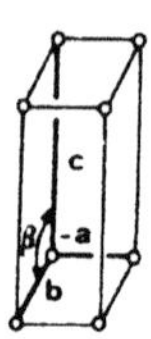	$c_{11} \quad c_{22} \quad c_{33}$ $0 \quad 0 \quad -\lvert c_{23}\rvert$	$\begin{vmatrix} 0 & 1 & 0 \\ -1 & 0 & 0 \\ 0 & 0 & 1 \end{vmatrix}$
P	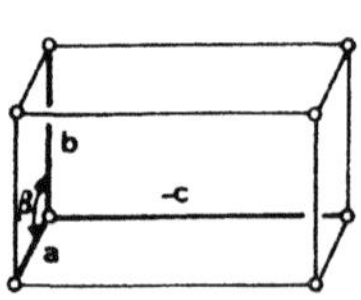	$c_{11} \quad c_{22} \quad c_{33}$ $-\lvert c_{12}\rvert \quad 0 \quad 0$	$\begin{vmatrix} 1 & 0 & 0 \\ 0 & 0 & -1 \\ 0 & 1 & 0 \end{vmatrix}$

Table 4.2. (continued)

Bravais-lattice	$\mathbb{Z}$-reduced form	transformation W
C	$\begin{matrix} c_{11} & c_{22} & c_{33} \\ -\frac{1}{2}c_{11} & 0 & -\lvert c_{23}\rvert \end{matrix}$	$\begin{vmatrix} -1 & -2 & 0 \\ -1 & 0 & 0 \\ 0 & 0 & -1 \end{vmatrix}$
C	$\begin{matrix} c_{11} & c_{22} & c_{33} \\ 0 & -\lvert c_{13}\rvert & -\frac{1}{2}c_{22} \end{matrix}$	$\begin{vmatrix} 0 & -1 & -2 \\ 0 & -1 & 0 \\ -1 & 0 & 0 \end{vmatrix}$
C	$\begin{matrix} c_{11} & c_{22} & c_{33} \\ 0 & -\frac{1}{2}c_{11} & -\lvert c_{23}\rvert \end{matrix}$	$\begin{vmatrix} 1 & 0 & 2 \\ 1 & 0 & 0 \\ 0 & 1 & 0 \end{vmatrix}$
C	$\begin{matrix} c_{11} & c_{11} & c_{33} \\ \lvert c_{12}\rvert & \lvert c_{13}\rvert & \lvert c_{13}\rvert \end{matrix}$	$\begin{vmatrix} -1 & -1 & 0 \\ 1 & -1 & 0 \\ 0 & 0 & 1 \end{vmatrix}$
C	$\begin{matrix} c_{11} & c_{11} & c_{33} \\ -\lvert c_{12}\rvert & -\lvert c_{13}\rvert & -c\lvert_{13}\rvert \end{matrix}$	$\begin{vmatrix} 1 & 1 & 0 \\ -1 & 1 & 0 \\ 0 & 0 & 1 \end{vmatrix}$
C	$\begin{matrix} c_{11} & c_{22} & c_{22} \\ \lvert c_{12}\rvert & \lvert c_{12}\rvert & \lvert c_{23}\rvert \end{matrix}$	$\begin{vmatrix} 0 & -1 & -1 \\ 0 & 1 & -1 \\ 1 & 0 & 0 \end{vmatrix}$
C	$\begin{matrix} c_{11} & c_{22} & c_{22} \\ -\lvert c_{12}\rvert & -\lvert c_{12}\rvert & -\lvert c_{23}\rvert \end{matrix}$	$\begin{vmatrix} 0 & 1 & 1 \\ 0 & -1 & 1 \\ 1 & 0 & 0 \end{vmatrix}$

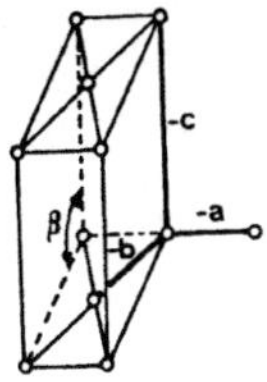

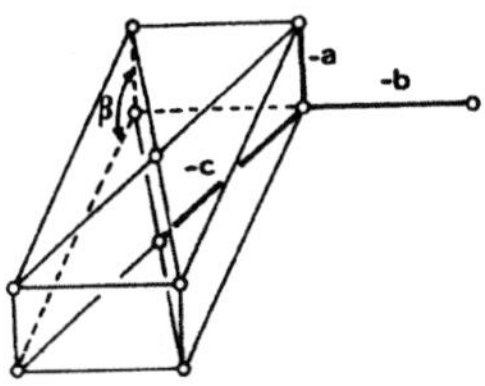

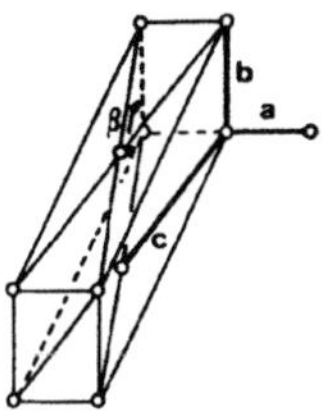

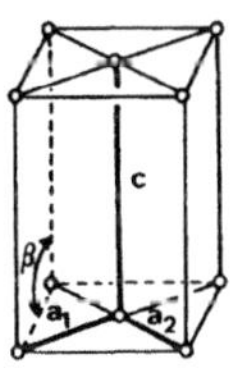

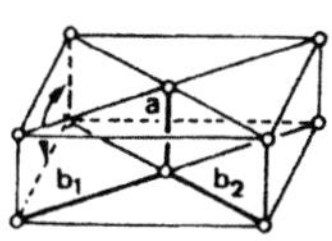

Table 4.2. (continued)

Bravais-lattice		$\mathbb{Z}$-reduced form	transformation W
C	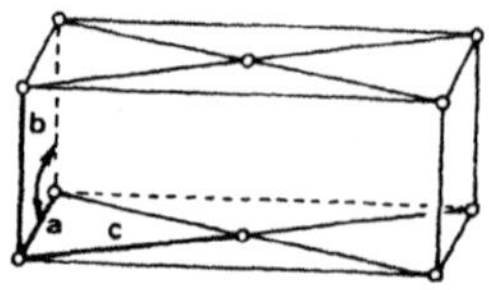	c_{11} c_{22} c_{33} $\lvert c_{12}\rvert$ $\frac{1}{2}c_{11}$ $\frac{1}{2}\lvert c_{12}\rvert$	$\begin{vmatrix} -1 & 0 & 0 \\ -1 & 0 & 2 \\ 0 & 1 & 0 \end{vmatrix}$
C	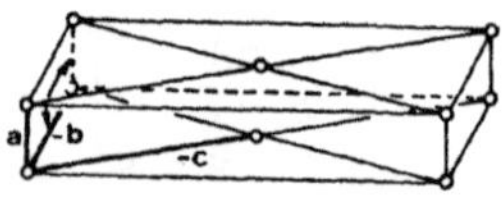	c_{11} c_{22} c_{33} $\lvert c_{12}\rvert$ $\frac{1}{2}\lvert c_{12}\rvert$ $\frac{1}{2}c_{22}$	$\begin{vmatrix} 0 & 1 & 0 \\ 0 & 1 & -2 \\ -1 & 0 & 0 \end{vmatrix}$
C	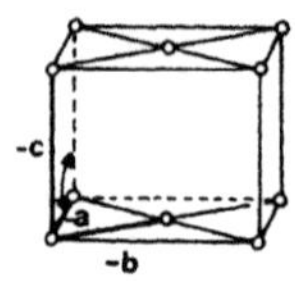	c_{11} c_{22} c_{33} $\frac{1}{2}c_{11}$ $\lvert c_{13}\rvert$ $\frac{1}{2}\lvert c_{13}\rvert$	$\begin{vmatrix} 1 & 0 & 0 \\ 1 & -2 & 0 \\ 0 & 0 & -1 \end{vmatrix}$
C	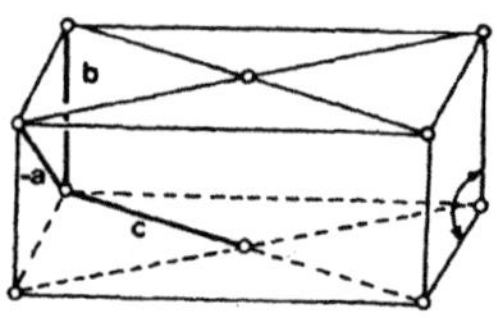	c_{11} c_{22} c_{33} $-\lvert c_{12}\rvert$ $\frac{1}{2}(\lvert c_{12}\rvert-c_{11})$ $\frac{1}{2}(\lvert c_{12}\rvert-c_{22})$	$\begin{vmatrix} -1 & -1 & 0 \\ 1 & 1 & 2 \\ 0 & 1 & 0 \end{vmatrix}$
C	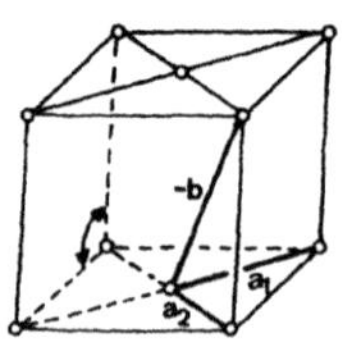	c_{11} c_{11} c_{33} $\lvert c_{13}\rvert+\lvert c_{23}\rvert-c_{11}$ $-\lvert c_{13}\rvert$ $-\lvert c_{23}\rvert$	$\begin{vmatrix} 0 & 1 & 1 \\ -1 & -1 & 0 \\ 1 & -1 & 0 \end{vmatrix}$
C	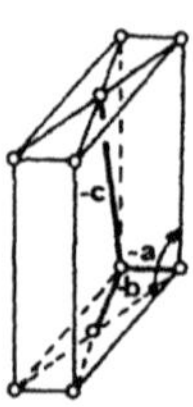	c_{11} c_{22} c_{33} $\frac{1}{2}c_{11}$ $\frac{1}{2}c_{11}$ $\lvert c_{23}\rvert$	$\begin{vmatrix} 0 & -1 & 1 \\ 1 & 0 & 0 \\ -1 & 2 & 0 \end{vmatrix}$

Table 4.2. (continued)

Bravais-lattice	$\mathbb{Z}$-reduced form	transformation W
7. <u>Triclinic crystal system</u>		
P 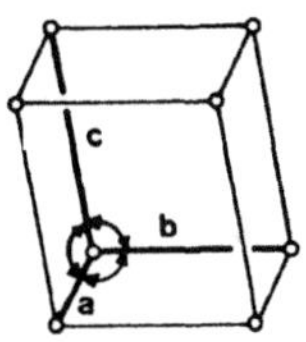	c_{11} c_{22} c_{33} $\lvert c_{12}\rvert$ $\lvert c_{13}\rvert$ $\lvert c_{23}\rvert$	$\begin{vmatrix} 1 & 0 & 0 \\ 0 & 1 & 0 \\ 0 & 0 & 1 \end{vmatrix}$
	c_{11} c_{22} c_{33} $-\lvert c_{12}\rvert$ $-\lvert c_{13}\rvert$ $-\lvert c_{23}\rvert$	$\begin{vmatrix} 1 & 0 & 0 \\ 0 & 1 & 0 \\ 0 & 0 & 1 \end{vmatrix}$

4.4. <u>The reduction scheme of Selling</u>

4.4.1. Reduction conditions

An alternative reduction scheme in E^3, which however is not conform to our definition 4.1, was described by Selling (Sell1). Selling introduced a fourth lattice vector

$$\vec{a}_4 := -\vec{a}_1-\vec{a}_2-\vec{a}_3,$$

so that the vector sum becomes zero,

$$\vec{a}_1+\vec{a}_2+\vec{a}_3+\vec{a}_4=0.$$

The four lattice vectors define six homogeneous coefficients c_{12}, c_{13}, c_{14}, c_{23}, c_{24}, and c_{34}.

<u>Definition 4.4:</u> The six homogeneous coefficients in E^3 are $\mathbb{Z}$-reduced according to Selling (also Delaunay-reduced) if

$$\sigma := |\vec{a}_1|^2+ |\vec{a}_2|^2+ |\vec{a}_3|^2+ |\vec{a}_4|^2$$

is minimal.

The minimum of σ is obtained if all six homogeneous coefficients are negative,

$$c_{12},\ c_{13},\ c_{14},\ c_{23},\ c_{24},\ c_{34} \le 0.$$

Indeed, if for example

$$c_{12} := |\vec{a}_1|\ |\vec{a}_2|\ \cos\alpha_{12} > 0$$

then we set

$$\vec{a}_1' := -\vec{a}_1,\ \vec{a}_2' := \vec{a}_2,$$

so that the angle between $\vec{a}_1$ and $\vec{a}_2$ becomes obtuse. In order to keep the vector sum zero we set

$$\vec{a}_3' := \vec{a}_3 + \vec{a}_1, \quad \vec{a}_4' := \vec{a}_4 + \vec{a}_1.$$

It follows that

$$\sigma' := |\vec{a}_1|^2 + |\vec{a}_2|^2 + |\vec{a}_3+\vec{a}_1|^2 + |\vec{a}_4+\vec{a}_1|^2$$

$$= \sigma + 2\vec{a}_1 \bullet (\vec{a}_1+\vec{a}_3+\vec{a}_4) = \sigma - 2\vec{a}_1 \bullet \vec{a}_2$$

$$= \sigma - 2c_{12}.$$

Hence, σ' is reduced by $2c_{12}$. The number of such reduction steps is finite because within a ball of radius $\sqrt{\sigma}$ only a finite number of lattice vectors exists. Such a minimum is asserted for every lattice.

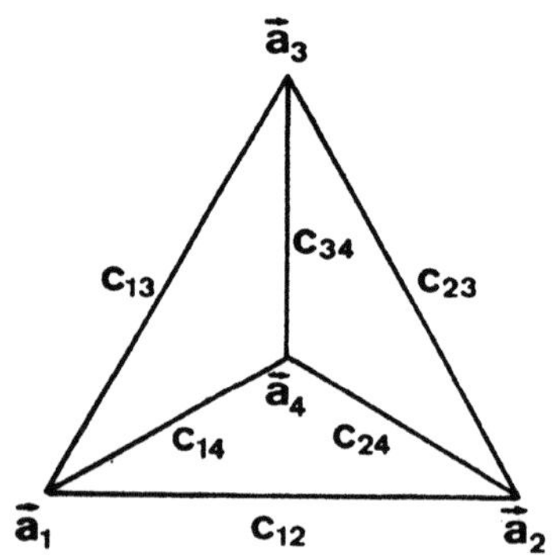

Figure 4.2. The Delaunay tetrahedron

4.4.2. Determination of the Selling reduced form

Delaunay (Delo2) has visualized such a reduction step on a tetrahedron as shown in Figure 4.2. The vertices of the tetrahedron are identified with the four lattice vectors $\vec{a}_h, \vec{a}_i, \vec{a}_j$, and $\vec{a}_k$. The edge connecting $\vec{a}_i$ with $\vec{a}_j$ is identified with the coefficient c_{ij}. If one of the six coefficients is positive, say $c_{ij}>0$ then a reduction step is necessary. The derived tetrahedron is obtained through the following transformations:

We set $\vec{a}'_i:=-\vec{a}_i$; $\vec{a}'_j:=\vec{a}_j$; $\vec{a}'_h:=\vec{a}_h+\vec{a}_i$; $\vec{a}'_k:=\vec{a}_k+\vec{a}_i$; Then the six derived coefficients become:

$$c'_{ij}:=-c_{ij};\quad c'_{jh}:=c_{jh}+c_{ij};\quad c'_{jk}:=c_{jk}+c_{ij};$$

$$c'_{ih}:=-\vec{a}_i\bullet(\vec{a}_h+\vec{a}_i)=-\vec{a}_i\bullet(-\vec{a}_j-\vec{a}_k)=\ c_{ij}+c_{ik};$$

$$c'_{ik}:=-\vec{a}_i\bullet(\vec{a}_k+\vec{a}_i)=-\vec{a}_i\bullet(-\vec{a}_j-\vec{a}_h)=\ c_{ij}+c_{ih};$$

$$c'_{hk}:=(\vec{a}_i+\vec{a}_h)\bullet(\vec{a}_i+\vec{a}_k)=(\vec{a}_i+\vec{a}_h)\bullet(-\vec{a}_j-\vec{a}_h)$$

$$=-\vec{a}_i\bullet\vec{a}_j-\vec{a}_i\bullet\vec{a}_h-\vec{a}_j\bullet\vec{a}_h+\vec{a}_h\bullet(\vec{a}_i+\vec{a}_j+\vec{a}_k)=\ c_{hk}-c_{ij}.$$

The results of this computations may be summerized in the following rules. If $c_{ij}>0$ then we proceed as follows:

1. Add c_{ij} to all coefficients whose edge meet the edge of c_{ij}.

2. Subtract c_{ij} from the coefficient whose edge does not meet the edge of c_{ij}.

3. Exchange the coefficients c_{ih} and c_{ik}.

4. Change the sign of c_{ij}.

Two lattice bases are equivalent if they agree in the six reduced homogeneous coefficients c_{ij}. On the other hand the six homogeneous coefficients do not determine a unique lattice basis. Patterson and Love (Patt1) gave examples where different lattice bases, belonging to the same lattice, result in the same reduced homogeneous coefficients c_{ij}.

The basis vectors $\vec{a}_1$, $\vec{a}_2$, and $\vec{a}_3$ determined from the six reduced homogeneous coefficients not always include the three shortest linearly independent lattice vectors. However, these are included amongst the lattice vectors

$$\vec{a}_1,\ \vec{a}_2,\ \vec{a}_3,\ \vec{a}_4,\ \vec{a}_1+\vec{a}_2,\ \vec{a}_1+\vec{a}_3,\ \vec{a}_2+\vec{a}_3.$$

Delaunay (Delo2) has described the 24 various kinds of Selling reduced lattice types.

4.5. The reduction scheme of Minkowski

Contrary to the two- and three-dimensional lattice, it is remarkable that in higher dimensional lattices n shortest linearly independent lattice vectors do not necessarily form a lattice basis. This is demonstrated in the following example due to Bieberbach and Schur (Bieb3).

Example 4.1: In E^5 consider the following vectors, referred to a cartesian coordinate system,

$$\vec{a}_1 := (1,0,0,0,0)$$
$$\vec{a}_2 := (0,1,0,0,0)$$
$$\vec{a}_3 := (0,0,1,0,0)$$
$$\vec{a}_4 := (0,0,0,1,0)$$
$$\vec{a}_5 := (\tfrac{1}{2},\tfrac{1}{2},\tfrac{1}{2},\tfrac{1}{2},\tfrac{1}{2}).$$

They form a lattice basis of a lattice T^5. The length of $\vec{a}_5$ is $\sqrt{5}/2$. We determine the lattice vector $\vec{b}\in T^5$,

$$\vec{b} := -\vec{a}_1-\vec{a}_2-\vec{a}_3-\vec{a}_4+2\vec{a}_5 = (0,0,0,0,1).$$

Although $\vec{b}$ is shorter than $\vec{a}_5$ it does not form together with $\vec{a}_1,\ldots,\vec{a}_4$ a lattice basis of T^5 because the volume of the corresponding parallelepiped is twice as large as that of $\vec{a}_1,\ldots,\vec{a}_5$. The basis vectors $\vec{a}_1,\ldots,\vec{a}_4$ generate a sublattice T^4. The cosets $\vec{t}_i+T^4$ generate an infinite stack of parallel lattice hyperplanes F_i which can be labeled in the following way (cf. section 3.4),

$$\ldots,F_{-2},\ F_{-1},\ F_0=T^4,\ F_1,\ F_2,\ldots$$

The basis vector $\vec{a}_5$ ends in F_1 whereas the shorter lattice vector $\vec{b}$ ends in the second lattice plane F_2. We encounter here for the first time the phenomenon that a shortest linearly independent lattice vector does not end in a lattice hyperplane which has the smallest interlayer spacing from the sublattice T^4.

A general reduction scheme applicable to any dimension was proposed by Minkowski (Mink1). The above example shows that a shortest lattice vector which is linearly independent of the $\vec{a}_1,\ldots,\vec{a}_{k-1}$ does not necessarily belongs to a lattice basis in the sublattice T^k. However, if we take a shortest lattice vector among all those lattice vectors $\vec{t}:=m_1\vec{a}_1,\ldots,m_n\vec{a}_n$ which form together with $\vec{a}_1,\ldots,\vec{a}_{k-1}$ a basis in the sublattice T^k then a $\mathbb{Z}$-reduced lattice basis is obtained. The following theorem is due to Minkowski.

<u>Theorem 4.4:</u> Let $\vec{a}_1,\ldots,\vec{a}_n$ be a lattice basis of T^n. The lattice vector $\vec{t}:=m_1\vec{a}_1+\ldots+m_n\vec{a}_n$ forms together with the lattice vectors $\vec{a}_1,\ldots,\vec{a}_{k-1}$, $1<k\leq n$, a lattice basis of the sublattice T^k if the greatest common divisor of $m_k,\ldots,m_n$, $\gcd(m_k,\ldots,m_n)$, is equal to 1.

Proof: The basis vectors $\vec{a}_1,\ldots,\vec{a}_{k-1}$ generate the sublattice T'. As $\gcd(m_k,\ldots,m_n)=1$ it follows that not all of the $m_k,\ldots,m_n$ can be zero. Therefore, $\vec{t}$ is linearly independent of $\vec{a}_1,\ldots,\vec{a}_{k-1}$. Further it follows that for every integer $|h|>1$ the vector $\vec{t}/h$ is not a lattice vector. Therefore, the cosets $m\vec{t}+T'$, $m\in\mathbb{Z}$, generate the sublattice T^k.

When setting up the reduction conditions Minkowski did not consider the special case that $\vec{a}_k$ is not uniquely determined. This is the case if the orthogonal projection of $\vec{a}_k$ onto T' falls on the boundary of the Dirichlet parallelotope P' of T'. Also Van der Waerden (Waer2) and Weber (Webe2) gave only the main conditions for the reduced form. In order to make the reduction scheme complete selection rules have to be appended.

<u>Definition 4.5:</u> A positive-definite n-nary quadratic form is $\mathbb{Z}$-reduced according to Minkowski if the following conditions are fulfilled:

Main conditions:

1. $c_{kk} \leq f(m_1,\ldots,m_n)$, $\gcd(m_k,\ldots,m_n)=1$, $m_i\in\mathbb{Z}$
2. $c_{kk+1}\geq 0$

We now give selection rules to make the $\mathbb{Z}$-reduced form unique.

Selection rules:
Let M_1 be the set of all lattice vectors $\vec{r}_{1i}$ of length $|\vec{a}_1|$. For each $\vec{r}_{1i}\in M_1$ we take all lattice vectors $\vec{r}_{2i}$, referred to a lattice basis $\vec{r}_{1i},\vec{a}_2,\ldots,\vec{a}_n$, having $\gcd(m_2,\ldots,m_n)=1$ and length $|\vec{a}_2|$, and with maximal $c_{1i2j}:=\vec{r}_{1i}\vec{r}_{2j}\cos\alpha_{1i2j}$. Let M_2 be the set of all sets

$$B_2:=\{\vec{r}_{1i},\vec{r}_{2j}\ |\ |\vec{r}_{1i}|=|\vec{a}_1|,|\vec{r}_{2j}|=|\vec{a}_2|,c_{1i2j}=\max\}.$$

For each $B_2\in M_2$ we take all lattice vectors $\vec{r}_{3i}$, referred to a lattice basis $\vec{r}_{1i},\vec{r}_{2j},\vec{a}_3,\ldots,\vec{a}_n$, having

$\gcd(m_3,\dots,m_n)=1$ and length $|\vec{a}_3|$, and with maximal c_{2i3j}. Among these we take all those with maximal c_{1i3j}. Continuing in this way we end up with a set M_n of $\mathbb{Z}$-reduced lattice bases B_n. By construction, all of them have the same metric tensor C hence, they are related through a symmetry operation of T^n.

Even for low dimensions n=5 or n=6 the number of sets B_k may become very large. Therefore, the a priori knowledge of the symmetry of the lattice T^n drastically can reduce the calculations envolved.

The above definition 4.5 contains an unlimited set of reduction conditions. Therefore, the first finiteness theorem of Minkowski (Mink1, Bieb3) is decisive.

Theorem 4.6: The reduction conditions according to Minkowski all arise from a finite number amongst them.

Proof: Let $\vec{t}:=m_1\vec{a}_1+\dots+m_n\vec{a}_n$, $\text{bdc}(m_k,\dots,m_n)=1$. Assume that $\vec{a}_1,\dots,\vec{a}_{k-1}$ is a $\mathbb{Z}$-reduced basis in the sublattice T' and let P' be its Dirichlet parallelotope. A shortest lattice vector $\vec{r}$ of the coset $\vec{t}+T'$ has an orthogonal projection onto T' which falls into the interior or on the boundary of P' because otherwise a shorter lattice vektor $\vec{r}'\in\vec{t}+T'$ would exist. It follows that the facets of P' give the necessary restrictions for $\vec{r}$. Let $\vec{f}:=h_1\vec{a}_1+\dots+h_{k-1}\vec{a}_{k-1}$ be a facet vector of P'. By theorem 2.2, the parallelotope P' has only a finite number of facets hence, a finite number of inequalities of the form

$$\vec{f}\cdot\vec{r}\le\tfrac{1}{2}|\vec{f}|$$

determine $\vec{r}$.

Minkowski proved that for $n \le 4$ it is sufficient to consider the values -1, 0, 1 for the m_i. This corresponds to the observation that up to T^3 only lattice vectors $\vec{f}=m_1\vec{a}_1+m_2\vec{a}_2+m_3\vec{a}_3$, $m_i=-1,0,1$, referred to a $\mathbb{Z}$-reduced lattice basis, are facet vectors. For the determination of $\vec{a}_4$ in T^4 the coefficient m_4 is 1 by definition.

In two- and three-dimensional space two respectively three shortest linearly independent lattice vectors form a lattice basis. It follows from theorem 3.8 that these basis vectors are facet vectors of the Dirichlet parallelotope. In higher dimensional spaces this is no longer true. Let $\vec{a}_1,\ldots,\vec{a}_n$ be a Minkowski $\mathbb{Z}$-reduced lattice basis in E^n. The lattice is T^n and P_0 is its parallelotope. The basis vectors $\vec{a}_1,\ldots,\vec{a}_{k-1}$ generate a sublattice $T'CT^n$ with parallelotope P'. Definition 4.5 only requires that the orthogonal projection of $\vec{a}_k$ onto the sublattice T' falls into, or on the boundary, of P'. However, $\vec{a}_k$ need not be a facet vector of P_0.

From definition 4.6 it follows immediately that

$$0 < c_{11} \le c_{22} \le \ldots \le c_{nn}.$$

However, as we have seen, the c_{ii} not necessarily represent shortest linearly independant lattice vectors.

The following inequality of Minkowski holds,

$$\lambda_n c_{11} c_{22} \ldots c_{nn} \le |\det(C)|,$$

with

$$\lambda_1 = 3/4, \quad \lambda_2 = 1/2, \quad \lambda_3 = 1/4.$$

Lower limits for the λ_n were given by Blichfeldt and Remak (Blic1,Rema1).

$$\lambda_n = \left|\frac{\pi}{2}\right|^n \left|\frac{1}{\Gamma(2+\frac{n}{2})}\right|^2 \qquad \text{for } n \le 4$$

$$\lambda_n = \left|\frac{\pi}{2}\right|^n \left|\frac{1}{\Gamma(2+\frac{n}{2})}\right|^2 \left|\frac{4}{5}\right|^{\frac{1}{2}(n-3)(n-4)} \qquad \text{for } n > 4,$$

where Γ is the gamma function.

4.6. Historical remarks

The mathematician Joseph Louis Lagrange (1736-1813) was the first to show that in each class of equivalent positive-definite binary quadratic forms a unique ℤ-reduced form exists which fulfils a set of inequalities (Lagr1). The binary and ternary quadratic forms were also investigated by the mathematician Carl Friedrich Gauss (1777-1855) in his "Disquisitiones arithmeticae". However, the difficult problem to extend Lagrange's theory to the ternary forms was achieved by the physicist Ludwig August Seeber in 1831 (Seeb2). Gauss had reviewed Seeber's book and has appriciated his results although he mentioned that the proofs were very lengthy and cumbersome (Gaus1). Seeber had stated the auxillary conditions in terms of the adjoint lattice which is called now the dual or reciprocal lattice. This is the first time that the reciprocal lattice is used in crystallography. The present formulation of the auxillary conditions is due to Gotthold Eisenstein (1823-1852) (Eise1). In 1850 the mathematician Peter Gustav Lejeune Dirichlet (1805-1859) gave a geometric interpretation of Seeber's reduction conditions through the construction of the Dirichlet parallelogon. The general case of the positive-definite n-nary quadratic forms was solved in 1905 by the mathematician Hermann Minkowski (1864-1909) (Mink1). Dirichlet's method was further developed by M. Georges Voronoĭ (1868-1908) who investigated the parallelotopes in higher dimensions.

In 1824 Seeber proposed to represent the atoms or molecules by a system of small spherical balls which form a space lattice (cf. section 1.6). With his reduction conditions he was able to classify the possible lattice types according to their ℤ-reduced forms. Seeber hoped that this classification would describe all possible crystal structures. However, Seeber's work on crystal lattices was largely overlooked by crystallographers at that time. Only in 1928 Niggli introduced the reduction theory of Seeber into crystallography and he completely enumerated the 44 types of ℤ-reduced forms (Nigg3). But again Niggli's work was overlooked for a long time. In 1932 Boris Nikolaevič Delaunay introduced the reduction scheme of Selling into crystallography and he developed a simple algorithm to obtain the Selling reduced homogeneous coefficients which is based on a representation of these coefficients in a tetrahedron (Delo2). However, the Selling reduction does not give a unique lattice basis. Because of this disadvantage simplified versions of Seeber's reduction conditions were again used to find a reduced lattice basis. In the years 1960 to 1978 many incomplete results were published in the crystallographic literature. It is due to Křivý and Gruber that the work of Niggli was rediscovered in 1978. However, these authers, as well as the previous ones, overlooked that already Seeber in 1831 gave a complete reduction algorithm.

Ludwig August Seeber was born in Karlsruhe (Germany) in 1793. Between the years 1819-1822 he was a teacher at the Military School in Karlsruhe. He was then appointed a professor in physics at the University of Freiburg (Germany). In the year 1834 he was appointed a professor in physics at the Polytechnical School in Karlsruhe. However, he was not very successful there and had severe problems of discipline so that in 1840 he had to retire in good time. He died in 1855.

5. Crystallographic Symmetry Operations

In this chapter we investigate the restrictions imposed by a lattice on the symmetry operations acting on a regular point system $X \subset E^n$.

5.1. Definitions

Let G be a space group acting transitively on a regular point system X. By theorem 3.2, the existence of a lattice $T^n < G$ is asserted,

$$T^n := \{\vec{t} \mid \vec{t} = m_1\vec{a}_1 + \ldots + m_n\vec{a}_n,\ m_i \in \mathbb{Z}\}.$$

The lattice vectors $\vec{a}_1, \ldots, \vec{a}_n$ form a lattice basis. In order to get the possible symmetry operations $(S, \vec{s}) \in G$ we have to determine within the Euclidean group $\mathbb{E}(n)$ of all isometries those symmetry operations which map the lattice T^n onto itself.

Definition 5.1: A crystallographic symmetry operation is an isometry which brings a lattice T^n into self-coincidence.

Every translation $(I, \vec{t})$ of T^n is a crystallographic symmetry operation. We take a conjugate of $(I, \vec{t})$ in G. For a symmetry operation $(S, \vec{s}) \in G$ we calculate

$$\begin{aligned}(S,\vec{s})(I,\vec{t})(S,\vec{s})^{-1} &= (S, S\vec{t}+\vec{s})(S^{-1}, -S^{-1}\vec{s}) \\ &= (I, -SS^{-1}\vec{s}+S\vec{t}+\vec{s}) \\ &= (I, S\vec{t}).\end{aligned}$$

We require that the symmetry operation $(I, S\vec{t})$ again is a translation of T^n. It follows that the rotation part S of the symmetry operation maps the lattice onto itself and thereby leaves the origin fixed. A necessary condition for a symmetry operation $(S, \vec{s})$ to have this property is given in the following theorem.

Theorem 5.1: The rotation part S of every crystallographic symmetry operation has a representation in E^n through a unimodular nxn matrix with integer coefficients.

Proof: The rotation S is an isometry therefore, the representation matrix is unimodular (cf. section 1.3). S carries the lattice basis $\vec{a}_1,\ldots,\vec{a}_n$ into the lattice vectors $\vec{t}_1,\ldots,\vec{t}_n$. In reference to the lattice basis $\vec{a}_1,\ldots,\vec{a}_n$, every lattice vector has integer components therefore, the representation matrix has integer coefficients s_{ij}.

Every rotation $S \in G$ maps a lattice vector $\vec{t}$ onto a new lattice vector $\vec{t}'$ of equal length $|\vec{t}|$. By condition 1.1 only a finite number of lattice vectors of length $|\vec{t}|$ exists within a ball of radius $|\vec{t}|$. Therefore, the original system of basis vectors has to be identically restored after a finite number of repetitions. If the integer k is the smallest such number then

$$S^k = I.$$

The integer k is called the order of the rotation S. We calculate

$$(S,\vec{s})^k = (I,S^{k-1}\vec{s}+S^{k-2}\vec{s}+\ldots+S\vec{s}+\vec{s}).$$

The screw vector $\vec{t}:=S^{k-1}\vec{s}+\ldots+S\vec{s}+\vec{s}$ again is a lattice vector of T^n, which poses restrictions on the shift vector $\vec{s}$. These restrictions will be considered later in sections 5.5 and 8.2.

The following definitions and results are standard in linear algebra. If for $\vec{x} \in E^n$, $\vec{x}\neq 0$, we have

$$S\vec{x} = \lambda\vec{x},$$

then $\vec{x}$ is the eigenvector and λ the eigenvalue of S. We rewrite this equation in the form

$$(S-\lambda I)\vec{x} = 0.$$

In order that λ is an eigenvalue of S it is necessary that

$$\det(S-\lambda I) = 0.$$

The determinant $\det(S-\lambda I)$ is called the characteristic polynom of S.

<u>Theorem 5.2:</u> The characteristic polynom $\det(S-\lambda I)$ is invariant under a transformation of the basis.

Proof: Under a change of basis S is transformed into A^tSA^o (cf. section 3.2). We recall that

$$\det(AS)=\det(A)\det(S),\ \det(A^{-1})=1/\det(A),\ \text{and}\ AI=IA.$$

Referred to the new basis, the characteristic polynom becomes

$$\det(A^tSA^o-\lambda I) = \det(A^tSA^o-\lambda A^tIA^o)$$
$$= \det(A^t(S-\lambda)A^o)$$
$$= \det(S-\lambda I).$$

The characteristic polynom is of degree n in λ,

$$\varphi_n\lambda^n+ \varphi_{n-1}\lambda^{n-1}+\ldots+ \varphi_1\lambda+ \varphi_0.$$

We note that φ_n is equal to the determinant det(S) and that φ_{n-1} is equal to the trace $tr(S):=\sum_{i=1}^{n}s_{ii}$. The invariance of the characteristic polynom requires the invariance of the coefficients $\varphi_n,\ldots,\varphi_0$. In particular det(S) and tr(S) are invariant under a change of basis.

5.2. Rotations in E^2

In the plane E^2 we consider a rotation S which carries the point x over to the point x'. In reference to a cartesian coordinate system, the rotation S can be represented through a 2x2 matrix

$$S^+ := \begin{vmatrix} \cos\varphi & -\sin\varphi \\ \sin\varphi & \cos\varphi \end{vmatrix}$$

if S is a proper rotation, or

$$S^- := \begin{vmatrix} -1 & 0 \\ 0 & 1 \end{vmatrix}$$

if S is a reflection.

S^+ is an anticlockwise rotation with rotation angle φ. The determinant of S^+ is

$$\det(S^+) = \cos^2\varphi + \sin^2\varphi = 1.$$

S^- is a reflection in the second coordinate axis. The determinant $\det(S)=-1$.

We now determine among all rotations in the plane those rotations which leave the lattice T^2 invariant. By theorem 5.1, these are just the rotations which allow an integral representation by 2x2 matrices. The characteristic ploy-nom of S becomes,

$$\det(S-\lambda I) = \lambda^2 - 2\lambda\cos\varphi + 1.$$

We require that S assumes an integral representation therefore, it follows from theorem 5.2 that the coefficients of the characteristic polynom have to be integers.

We calculate the trace of S^+ and set it equal to an integer p.

$$\mathrm{tr}(S^+) = 2\cos\varphi = p.$$

From this equation follow the allowed rotation angles φ for the crystallographic rotations in the plane E^2. The results are shown in Table 5.1.

Table 5.1. Allowed rotation angles φ for crystallographic rotations S^+ in the plane E^2

φ	k	p
0°	1	2
60°	6	1
90°	4	0
120°	3	-1
180°	2	-2

In the plane E^2 only the crystallographic rotations S^+ of order 1, 2, 3, 4 and 6 are allowed. This is the crystallographic restriction imposed by the lattice T^2.

5.3. Rotations in E^n

The procedure described in section 5.2. can be applied to spaces of any finite dimension. In order to get an insight into the rotations in E^n the theorem of Schläfli (Schl2) is of importance.

Theorem 5.3: Every rotation S in E^n can be represented through rotations in a set of mutually perpendicular one- and two-dimensional subspaces.

Proof: The following proof is due to Buser (Buse1). In order to show that E^n, under the action of S, splits into invariant one- or two-dimensional subspaces we use the property that a point x on the unit sphere covers a minimal distance under the rotation S if it is contained in an invariant subspace. We choose a unit vector $\vec{x}\in E^n$ for which the angle $\varphi:=\sphericalangle(\vec{x},S\vec{x})$ is minimal. If $\varphi=0°$ or $\varphi=180°$ then $\vec{x}$ determines a one-dimensional invariant subspace $U\subset E^n$. We take the orthogonal complement $E^{\perp}$ and determine again a vector $\vec{x}\in E^{\perp}$ such that the angle $\varphi:=\sphericalangle(\vec{x},S\vec{x})$ is minimal. Suppose now that $\varphi\neq 0°$ or $\varphi\neq 180°$. We have to show that $\vec{x}$ and $S\vec{x}$ span an invariant two-dimensional subspace U'. We take any other unit vector $\vec{y}\in U'$ as shown in Figure 5.1. Let $\varphi':=\sphericalangle(\vec{y},S\vec{y})$ be the angle between $\vec{y}$ and $S\vec{y}$. Because S is an isometry it follows that $\delta:=\sphericalangle(\vec{x},\vec{y}) =\sphericalangle(S\vec{x},S\vec{y})$. If $\vec{y}$ lies between $\vec{x}$ and $S\vec{x}$ then the triangle inequality implies

$$\varphi'\leq(\varphi-\delta)+\delta = \varphi.$$

As φ is minimal it follows that $\varphi'=\varphi$ and therefore, $S\vec{y}$ has to lie in the subspace U'. Otherwise if $\vec{y}$ does not lie between $\vec{x}$ and $S\vec{x}$ then we take $\varphi:=180°-\varphi$, $\varphi':=180°-\varphi'$, and $\delta:=180°-\delta$. The triangle inequality then implies

$$\varphi'\geq(\varphi-\delta)+\delta = \varphi.$$

As φ is now maximal it follows that $\varphi'=\varphi$ and therefore, $S\vec{y}\in U'$. Hence, $U'\subset E^{\perp}$ is an invariant subspace. Repeating this step we see that E^n can be completely decomposed.

The two-dimensional invariant subspaces were called principal planes by Schläfli.

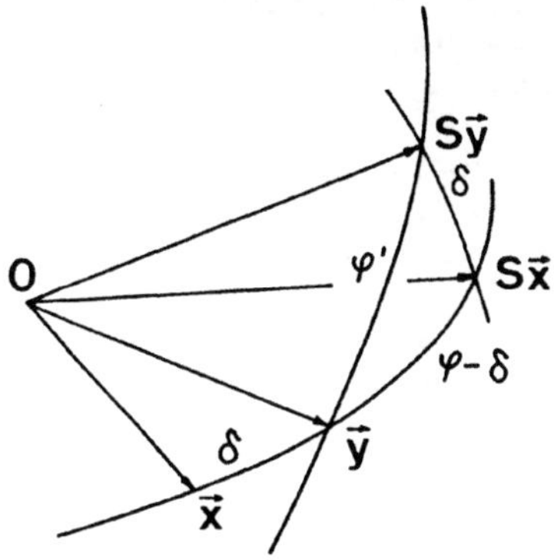

Figure 5.1

Referred to a cartesian coordinate system assumed within these invariant subspaces the matrix representation of a rotation S has a block-diagonal structure:

$$S := \begin{pmatrix} I_h & & & & & \\ & -I_k & & & 0 & \\ & & S_1 & & & \\ & & & \cdot & & \\ & 0 & & & \cdot & \\ & & & & & S_m \end{pmatrix} ,$$

where I_h, I_k are identity matrices of dimension h and k respectively and S_i, $i=1,\dots,m$, is a proper rotation of the form

$$S_i := \begin{vmatrix} \cos\varphi_i & -\sin\varphi_i \\ \sin\varphi_i & \cos\varphi_i \end{vmatrix} .$$

Each S_i represents in the principal plane π_i a rotation through the angle φ_i.

The decomposition of E^n into principal planes is not unique. If two or more among the angles φ_i are equal then the linear combination of any two vectors in the corresponding principal planes is again an invariant plane. However, the parts I_h, $-I_k$ are unique and the corresponding subspaces are mutually orthogonal and both are orthogonal to the principal planes π_i.

For the determination of those rotations which leave a lattice T^n invariant we have to introduce the concept of $\mathbb{Z}$-irreducible matrices.

The rotations S^m, $m=1,\dots,k$, are represented through unimodular nxn matrices. These matrices are called $\mathbb{Z}$-reducible if every matrix S^m can be represented in the same block-diagonal form, where all submatrices W_i, $i=1,\dots,v$, are unimodular with integer coefficients, and the remaining coefficients outside of these blocks are zero.

$$S^m = \begin{pmatrix} W_1 & & & & \\ & W_2 & & 0 & \\ & & \cdot & & \\ & 0 & & \cdot & \\ & & & & W_v \end{pmatrix}, \quad m=1,\dots,k.$$

Each submatrix W_i, $i=1,\dots,v$, maps the corresponding sublattice T_i of dimension n_i onto itself. We require that these submatrices have integer coefficients hence, they are referred to a lattice basis.

If no such representation exists simultaneously for all rotation matrices S^m, $m=1,\dots,k$, then these matrices are called $\mathbb{Z}$-irreducible. $\mathbb{Z}$-irreducible matrices map the lattice T^n onto itself and leave no proper sublattice invariant.

Example 5.1: All matrices representing the rotations of order $k=3$ in the lattice T^3 can be brought into the following $\mathbb{Z}$-reduced form:

$$S^1 := \left(\begin{array}{cc|c} 0 & -1 & 0 \\ 1 & -1 & 0 \\ \hline 0 & 0 & 1 \end{array}\right) \quad S^2 := \left(\begin{array}{cc|c} -1 & 1 & 0 \\ -1 & 0 & 0 \\ \hline 0 & 0 & 1 \end{array}\right) \quad S^3 := \left(\begin{array}{cc|c} 1 & 0 & 0 \\ 0 & 1 & 0 \\ \hline 0 & 0 & 1 \end{array}\right).$$

In this case the lattice T^3 is decomposed into two mutual orthogonal invariant sublattices of dimensions two and one respectively.

In the lattice T^4 the following matrices representing the rotations of order $k=5$ are $\mathbb{Z}$-irreducible.

$$S^1 := \begin{pmatrix} 0 & -1 & 1 & 0 \\ 0 & -1 & 0 & 1 \\ 0 & -1 & 0 & 0 \\ 1 & -1 & 0 & 0 \end{pmatrix} \quad S^2 := \begin{pmatrix} 0 & 0 & 0 & -1 \\ 1 & 0 & 0 & -1 \\ 0 & 1 & 0 & -1 \\ 0 & 0 & 1 & -1 \end{pmatrix} \quad S^3 := \begin{pmatrix} -1 & 1 & 0 & 0 \\ -1 & 0 & 1 & 0 \\ -1 & 0 & 0 & 1 \\ -1 & 0 & 0 & 0 \end{pmatrix}$$

$$S^4 := \begin{pmatrix} 0 & 0 & -1 & 1 \\ 0 & 0 & -1 & 0 \\ 1 & 0 & -1 & 0 \\ 0 & 1 & -1 & 0 \end{pmatrix} \quad S^5 := \begin{pmatrix} 1 & 0 & 0 & 0 \\ 0 & 1 & 0 & 0 \\ 0 & 0 & 1 & 0 \\ 0 & 0 & 0 & 1 \end{pmatrix}.$$

In this case there exists no sublattice $T' \subset T^4$ which remains invariant. We note that in general, the principal planes are not lattice planes.

For a given lattice T^n the allowed orders of a crystallographic rotation S are given by the following theorem due to Hermann (Herm3):

Theorem 5.4: A rotation S of order k with integral matrix representation exists only in a lattice T^n of dimension $n \geq \psi(k)$.

Proof: Assume that the representation of S is $\mathbb{Z}$-irreducible. Then the rotation angles φ_i are all of the form $\varphi_i := 2\pi l_i/k$. By theorem 5.2, the coefficients of the characteristic polynom

$$(\lambda^2 - 2\lambda\cos 2\pi l_1/k + 1)\ldots(\lambda^2 - 2\lambda\cos 2\pi l_s/k + 1), \quad s := \tfrac{1}{2}n,$$

have to be integers. From analysis it is known that a solution only exists if $n = \psi(k)$.

The number-theoretic Euler totient function $\psi(k)$ gives the number of integers less than and relatively prime to k. An integer i is called relatively prime to k if gcd(i,k)=1. Alternatively, if k is decomposed into a product of prime factors

$$k = p_1^{h_1} p_2^{h_2} \ldots$$

then

$$\psi(k) := k \prod_{p_i | k} (1 - 1/p_i),$$

where the product extends over all different proper prime factors $p_i | k$ of k (for a proof see Hill1).

The trace of S,

$$tr(S) = 2\cos 2\pi l_1/k + \ldots + 2\cos 2\pi l_s/k, \quad s:=\tfrac{1}{2}n,$$

must be integral. This is fulfilled if the l_i are less than k/2 and are relatively prime to k. There exists exactly s such integers (using $k-l_i$ instead of l_i gives the same rotation because no absolute orientation of the space is assumed). Therefore, the integers l_i are uniquely determined for a crystallographic symmetry operation.

The formula for the Euler totient function reveals that it assumes even values if k>2. From this another theorem, due to Schläfli (Schl2), follows:

<u>Theorem 5.5:</u> The representations of crystallographic rotations S, which leave a lattice T^n of odd dimension n>1 invariant, are $\mathbb{Z}$-reducible.

Proof: For k>2 the Euler totient function $\psi(k)>0$ assumes only even values. This proves the theorem 5.5.

Table 5.2. Values of the Euler totient function $\psi(k)$

k	0	1	2	3	4	5	6	7	8	9
0	0	1	1	2	2	4	2	6	4	6
10	4	10	4	12	6	8	8	16	6	18
20	8	12	10	22	8	20	12	18	12	28
30	8	30	16	20	16	24	12	36	18	24
40	16	40	12	42	20	24	22	46	16	42
50	20	32	24	52	18	40	24	36	28	58
60	16	60	30	36	32	48	20	66	32	44
70	24	70	24	72	36	40	36	60	24	78
80	32	54	40	82	24	64	42	56	40	88
90	24	72	44	60	46	72	32	96	42	60

Table 5.2 can be inverted up to dimension 30 as is done in Table 5.3. Remarkably among these the dimensions 14 and 26 have no $\mathbb{Z}$-irreducible representations.

For n=2 we have the known result of section 5.2 that only rotations of order 1, 2, 3, 4, and 6 are allowed.

The chirality of a rotation in E^n (cf. section 1.4) is reveald by the following theorem due to Hermann (Herm3):

Table 5.3. Orders k of the $\mathbb{Z}$-irreducible representations for each dimension n up to 30

n	allowed orders	n	allowed orders
1	1, 2	16	17, 32, 34, 40, 48, 60
2	3, 4, 6	18	19, 27, 54
4	5, 8, 10, 12	20	25, 33, 50, 44, 66
6	7, 9, 14, 18	22	23, 46
8	15, 16, 20, 24, 30	24	35, 39, 45, 52, 56, 70, 78, 84, 90
10	11, 22	26	-
12	13, 21, 26, 28, 36, 42	28	29, 58
14	-	30	31, 62

Theorem 5.6: Let the matrix representation of a crystallographic rotation S be $\mathbb{Z}$-reduced. Then if the number of submatrices of order 2 is even the rotation S is a proper rotation, otherwise it is an improper rotation.

Proof: By assumption the matrix representation of S is $\mathbb{Z}$-reduced. The determinant det(S) is the product of the determinants of the individual submatrices in each block. $\mathbb{Z}$-irreducible submatrices of even dimension $n>1$ always represent proper rotations with positive determinant. The one-dimensional submatrices representing rotations of order 2 have negative determinant.

This allows us to find the allowed crystallographic rotations in lattices T^n of any finite dimension. First, all different decompositions of T^n into sublattices T_i with dimensions n_i are determined where all $n_i>1$ are even. Second, from Table 5.3 we find, for every sublattice T_i, the allowed orders of the rotation matrices. The order of the rotation S is then the smallest common multiple of the orders of all its submatrices. In Table 5.4 are shown the permissible rotations of the three-dimensional lattices. Each rotation is represented with respect to its sublattice decomposition. The rotations are arranged according to their orders k. Each small box represents a sublattice. The dimension n_i for each sublattice is not explicitely stated but with the help of Table 5.2 it can easily be deduced from the order of the submatrices W_i given in each box. If the rotation order 2 occurs in an odd number of times the rotation is an improper rotation. This is marked with an "I" in Tables 5.4 to 5.6. We note that in spaces of odd dimension the numbers of proper and improper rotations are equal.

Table 5.4. The crystallographic rotations S of the three-dimensional lattices T^3

k	crystallographic rotations		
1	[1|1|1|1]		
2	[2|1|1|1] I	[2|2|1|1]	[2|2|2|2] I
3	[3|1]		
4	[4|1]	[4|2] I	
6	[6|1]	[6|2] I	[3|2] I

In Table 5.5 the allowed crystallographic rotations S of the three-dimensional lattices T^3 are listed together with the usual crystallographic notation according to the International Tables, Vol. A.

Table 5.5. The Schönflies and the international notation for the crystallographic rotations of T^3

k	S	Schönflies notation	international notation
1	[1|1|1|1]	C_1	1
2	[2|1|1|1] I	$S_1=\sigma$	$\bar{2}$=m
2	[2|2|1|1]	C_2	2
2	[2|2|2|2] I	$S_2=i$	$\bar{1}$
3	[3|1]	C_3	3
6	[3|2] I	S_3	$\bar{6}$
4	[4|1]	C_4	4
4	[4|2] I	S_4	$\bar{4}$
6	[6|1]	C_6	6
6	[6|2] I	S_6	$\bar{3}$

In three-dimensional space the improper rotations may either be visualized as rotoreflections or else as rotoinversions. The rotoreflection S_k is a rotation with rotation angle $\varphi = 360°/k$ followed by a reflection in a mirror plane perpendicular to the rotation axis. This is the appropriate description of the improper rotation if we use the decomposition into invariant sublattices as we have done here. The rotoinversion $\bar{k}$ is a rotation with rotation angle φ= 360°/k followed by an inversion in the inversion center. Both notations are contrasted in Table 5.5.

In lattices of higher dimensions the rotations are more complicated and may not be visualized as easily as in three-dimensional space. In Table 5.6 are shown the permissible crystallographic rotions of the four-dimensional lattices.

Table 5.6. The crystallographic rotations S of the four-dimensional lattices T^4

k	crystallographic rotations				
1	[1|1|1|1]				
2	[2|1|1|1] I	[2|2|1|1]	[2|2|2|1] I	[2|2|2|2]	
3	[3|1|1]	[3|3]			
4	[4|1|1]	[4|2|1] I	[4|2|2]	[4|4]	
5	[5]				
6	[3|2|1] I	[3|2|2]	[6|2|1] I	[6|3]	[6|6]
8	[8]				
10	[10]				
12	[12]	[4|3]	[6|4]		

Example 5.2: The rotation [4|3] of order 12 leaves two two-dimensional sublattices invariant. This rotation can be visualized as a rotation by 90° in one sublattice and a simultaneous rotation by 120° in the other sublattice. Because both rotation angles are different the sublattices are mulual orthogonal. The rotation [5] of order 5 leaves no sublattice invariant. However, we may visualize the rotation in the two principal planes. An integral matrix representation of this rotation was given in Example 5.1.

From representation theory it is known that for a group H a projection operator P_v which projects the lattice basis $\vec{a}_1,\ldots,\vec{a}_n$ into the principal plane π_v is obtained through the formula:

$$P_v = \frac{1}{k}\sum_{i=1}^{k} \chi_v(R_i)R_i ,$$

where $\chi_v(R_i)$ is the character of the irreducible representations Γ_v, $R_i := (S_i^{-1})^t$, $S_i \in H$, is the transformation of the basis vectors $\vec{a}_1, \ldots, \vec{a}_n$, and k is the order of the group H. (for details the reader is referred to Brud1).

Example 5.3: For the rotation [5] we obtain two projection operators P_1 and P_2 corresponding to the two symmetry types of irreducible representations.

$$P_1 = \begin{pmatrix} b & 0 & c & c \\ -c & a & -c & 0 \\ 0 & -c & a & -c \\ c & c & 0 & b \end{pmatrix}, \qquad P_2 = \begin{pmatrix} a & 0 & -c & -c \\ c & b & c & 0 \\ 0 & c & b & c \\ -c & -c & 0 & a \end{pmatrix},$$

where $a := 1-\cos 72°$, $b := 1-\cos 144°$, $c := \cos 72° - \cos 144°$. The projection of the basis vectors $\vec{a}_1, \ldots, \vec{a}_4$ into the first principal plane π_1 gives the vectors $\vec{u}_1, \ldots, \vec{u}_4$. The vectors $\vec{u}_3$ and $\vec{u}_4$ are linearly dependent of $\vec{u}_1$ and $\vec{u}_2$. Thus we find for the first principel plane π_1

$$\vec{u}_1 = b\vec{a}_1 + c\vec{a}_3 + c\vec{a}_4,$$

$$\vec{u}_2 = -c\vec{a}_1 + a\vec{a}_2 - c\vec{a}_3.$$

Similarly, we have for the second principal plane π_2

$$\vec{v}_1 = a\vec{a}_1 - c\vec{a}_3 - c\vec{a}_4,$$

$$\vec{v}_2 = c\vec{a}_1 + b\vec{a}_2 + c\vec{a}_3.$$

The principal planes corresponding to rotations of order k=5, or k>6 are not lattice planes. Therefore, these principal planes contain besides the origin no other lattice points. Only for rotations of order 3, 4, and 6 the principal planes are sublattices too.

Referred to a cartesian coordiante system within the principal planes, the rotation of order 5 has the following representation:

$$S := \begin{pmatrix} \cos 72° & -\sin 72° & 0 & 0 \\ \sin 72° & \cos 72° & 0 & 0 \\ & & & \\ 0 & 0 & \cos 144° & -\sin 144° \\ 0 & 0 & \sin 144° & \cos 144° \end{pmatrix}.$$

The trace is easily calculated to be

$$\mathrm{tr}(S) = 2\cos 72° + 2\cos 144° = -1.$$

The rotation may be visualized by two simultaneous rotations in two mutual orthogonal principal planes. The first principal plane is rotated in itself through the angle of 72° leaving the second principal plane invariant. The second principal plane is rotated in itself through the angle of 144° leaving the first principal plane invariant. Only after repeating the rotation five times is the original status of both principal planes restored.

5.4. Symmetry support

A rotation S in E^n leaves at least the origin 0 fixed. We obtain the set of all fixed points of S from the following equation,

$$Sx = x, \; x \in E^n.$$

This equation is written in the form

$$(S-I)x = 0.$$

The solutions of this equation form a subspace $U \subset E^n$ of dimension $d=n-m$, where m is the rank of the matrix $(S-I)$. Only for the identity I, $d=n$ holds.

Definition 5.2: The subspace U which remains fixed under the action of the rotation S is called the symmetry support of S.

The symmetry support together with the coefficients of the characteristic polynom are important invariants which help to identify symmetry operations.

In what follows we give a practical method to calculate the symmetry support of S. We set $A:=S-I$. The equation $Ax=0$ can be written in form of a scheme as follows:

	μ_1	$\dots$	μ_n
ξ_1	a_{11}	$\dots$	a_{1n}
$\vdots$	$\vdots$		$\vdots$
ξ_n	a_{n1}	$\dots$	a_{nn}

If $a_{ij} \neq 0$ then ξ_i can be exchanged by μ_j using the algorithm of Gauss (for details the reader is referred to Nef1). A new scheme is calculated as follows,

$$\begin{array}{c|ccccc} & \mu_1 & \dots & \xi_i & \dots & \mu_n \\ \hline \xi_1 & b_{11} & \dots & b_{1j} & \dots & b_{1n} \\ \cdot & \cdot & & \cdot & & \cdot \\ \cdot & \cdot & & \cdot & & \cdot \\ \mu_j & b_{i1} & \dots & b_{ij} & \dots & b_{in} \\ \cdot & \cdot & & \cdot & & \cdot \\ \cdot & \cdot & & \cdot & & \cdot \\ \xi_n & b_{n1} & \dots & b_{nj} & \dots & b_{nn} \end{array}$$

where $b_{ij} := 1/a_{ij}$, $b_{ik} := -a_{ik}/a_{ij}$, $k \neq j$, $b_{hj} := a_{hj}/a_{ij}$, $h \neq i$, and $b_{hk} := a_{hk} - a_{hj}a_{ik}/a_{ij}$, $h,k \neq i,j$.

Let us assume that $\mu_1, \dots, \mu_m$ can be exchanged. Then after m steps we obtain the following scheme (possibly after renumbering)

$$\begin{array}{c|c|c|} & \xi_h \dots \xi_k & \mu_{m+1} \dots \mu_n \\ \hline \begin{array}{c}\mu_1 \\ \cdot \\ \cdot \\ \cdot \\ \mu_m\end{array} & P & Q \\ \hline \begin{array}{c}\xi_i \\ \cdot \\ \cdot \\ \cdot \\ \xi_j\end{array} & R & 0 \\ \hline \end{array}$$

The general solution is now given by

$$\begin{pmatrix} \mu_1 \\ \cdot \\ \cdot \\ \cdot \\ \mu_m \end{pmatrix} = Q \begin{pmatrix} \mu_{m+1} \\ \cdot \\ \cdot \\ \cdot \\ \mu_n \end{pmatrix}.$$

We note that only the matrix Q has to be calculated therefore, after exchanging ξ_i by μ_j the j-th column can be dropped. By arbitrarely asigning d real values for the $\mu_{m+1}, \dots, \mu_n$ we can calculate μ_1 to μ_m. Thus the fixed point f_1 is obtained,

$$\vec{f}_1 = \mu_1\vec{a}_1 + \ldots + \mu_n\vec{a}_n.$$

Besides the origin 0 we can find other d fixed points by choosing suitable real values for the $\mu_{m+1}, \ldots, \mu_n$ such that the linear hull of $0, f_1, \ldots, f_d$ span the subspace U.

Example 5.4: Let S be a crystallographic rotation in E^3 with the following matrix representation

$$\begin{vmatrix} 0 & 0 & -1 \\ 1 & 0 & 0 \\ 0 & -1 & 0 \end{vmatrix}$$

We calculate (S-I) and set up the following scheme:

	μ_1	μ_2	μ_3
ξ_1	-1	0	-1
ξ_2	1	-1	0
ξ_3	0	-1	-1

We exchange the ξ_i by the μ_j. The following steps are performed:

Step 1

	μ_2	μ_3
μ_1	0	-1
ξ_2	-1	-1
ξ_3	-1	-1

Step 2

	μ_3
μ_1	-1
μ_2	-1
ξ_3	0

Two of the ξ_i could be exchanged. Thus the rank is $m=2$ hence, the dimension of the symmetry support U is $d=3-2=1$. We arbitrarely assign $\mu_3=-1$ and calculate $\mu_1=1$, $\mu_2=1$. The fixed point f_1 becomes

$$\vec{f}_1 = 1\vec{a}_1 + 1\vec{a}_2 - 1\vec{a}_3.$$

It is easily verified that $Sf_1=f_1$. The rotation S leaves a one-dimensional subspace U fixed,

$$U := \{\lambda\vec{f}_1 \mid \lambda \in \mathbb{R}\}.$$

The rotation S has order 3. Hence, the symmetry support of S is the rotation axis [1,1,-1].

5.5. General symmetry operations in E^n

We now consider the general symmetry operation $(S,\vec{s})$ in E^n. The shift vector $\vec{s}$ depends on the choice of origin 0. Suppose that $0':=0+\vec{v}$ is an alternative origin. We calculate the transformation of $(S,\vec{s})$ under a change of origin. The symmetry operation $(S,\vec{s})$ carries the point x into y,

$$y = (S,\vec{s})x.$$

In reference to the new origin 0', the coordinates of the point x transform according to

$$x= x'+\vec{v}.$$

and therefore,

$$y'= Sx'+S\vec{v}-\vec{v}+\vec{s} = (S,S\vec{v}-\vec{v}+\vec{s})x'.$$

Thus the shift vector $\vec{s}'$, referred to the new origin 0', is

$$\vec{s}'= S\vec{v}-\vec{v}+\vec{s}.$$

In what follows we determine properties of $(S,\vec{s})$ which are independent of the choice of origin.

Let k be the order of S. In section 5.1 we calculated the screw vector

$$\vec{t}:= S^{k-1}\vec{s}+...+S\vec{s}+\vec{s}.$$

Theorem 5.7: The screw vector $\vec{t}$ is invariant under a change of origin.

Proof: In reference to a new origin $0':=0+\vec{v}$ the shift vector $\vec{s}$ is transformed according to

$$\vec{s}'= S\vec{v}-\vec{v}+\vec{s}.$$

We calculate the screw vector $\vec{t}'$ with respect to the new origin 0',

$$\vec{t}'=S^{k-1}\vec{s}'+\ldots+S\vec{s}'+\vec{s}'$$

$$= \vec{v}-S^{k-1}\vec{v}+S^{k-1}\vec{s}+\ldots+S^2\vec{v}-S\vec{v}+S\vec{s}+S\vec{v}-\vec{v}+\vec{s}$$

$$= S^{k-1}\vec{s}+\ldots+S\vec{s}+\vec{s} = \vec{t}.$$

Thus $\vec{t}$ has the required property.

By theorem 5.1, the rotation part S of a crystallographic symmetry operation has an integral representation, if referred to a lattic basis. In the following theorem the corresponding restriction for the shift vector $\vec{s}$ is revealed.

Theorem 5.8: If k is the order of the rotation S of a crystallographic symmetry operation $(S,\vec{s})$ then there exists a representation of $(S,\vec{s})$ in E^n such that the shift vector $\vec{s}$ has rational components with denominator k.

Proof: By theorem 3.1, the rotation S generates a cyclic group of finite order. Let the d-dimensional subspace $U \subset E^n$, $0 \leq d < n$, be the symmetry support of S. We can decompose E^n into U and its orthogonal complement $U^\perp$. With respect to U and $U^\perp$ the shift vector $\vec{s}$ has a unique decomposition

$$\vec{s} = \vec{s}'' + \vec{s}^\perp,$$

where $\vec{s}''$ and $\vec{s}^\perp$ are the orthogonal projections of $\vec{s}$ into U and $U^\perp$ respectively. Referred to this decomposition the rotation S has a representation in block-diagonal form

$$\begin{vmatrix} I_d & 0 \\ 0 & S^\perp \end{vmatrix}.$$

We determine the fixed point $\vec{p}^\perp \in U^\perp$ of the symmetry operation $(S^\perp,\vec{s}^\perp)$ by setting

$$\vec{p}^\perp = (S^\perp,\vec{s}^\perp)\vec{p}^\perp = S^\perp\vec{p}^\perp+\vec{s}^\perp.$$

It follows that

$$(I-S^\perp)\vec{p}^\perp = \vec{s}^\perp.$$

By construction of $U^\perp$ the rank of $(I-S^\perp)$ is $m=n-d$ therefore, this equation has a unique solution

$$\vec{p}^{\perp} = (I - S^{\perp})^{-1}\vec{s}^{\perp}.$$

We now choose the alternative origin $0' := 0 + \vec{p}^{\perp}$. With respect to the new origin the shift vector $\vec{s} = \vec{s}'' + \vec{s}^{\perp}$ becomes

$$\vec{s}' = S\vec{p}^{\perp} - \vec{p}^{\perp} + \vec{s}$$

$$= \vec{s}'' + (S^{\perp} - I)\vec{p}^{\perp} + \vec{s}^{\perp} = \vec{s}''.$$

Therefore, the symmetry operation $(S, \vec{s}')$ with respect to the new origin $0'$ is

$$(S, \vec{s}') = (S, \vec{s}'').$$

Because $\vec{s}''$ remains fixed under the rotation S the screw vector $\vec{t}'$ is,

$$\vec{t}' = k\vec{s}''.$$

It follows that $\vec{s}'' = \vec{t}'/k$. By left multiplication of $(S, \vec{s}'')$ with an appropriate translation, $(I, \vec{v})(S, \vec{s}'')$, we can always find a symmetry operation $(S, \vec{r}'')$ where

$$\vec{r}'' = (h_1\vec{a}_1 + \ldots + h_n\vec{a}_n)/k \quad \text{with } 0 \le h_i < k.$$

The following definition generalizes the definition 5.2. Let U be the symmetry support of the rotation S and $U^{\perp}$ its orthogonal complement. With respect to U and $U^{\perp}$ we have the following decomposition of the representation of $(S, \vec{s})$,

$$\begin{vmatrix} I_d & 0 \\ 0 & S^{\perp} \end{vmatrix}, \begin{vmatrix} \vec{s}'' \\ \vec{s}^{\perp} \end{vmatrix}.$$

With $\vec{p}^{\perp}$ we designate the fixed point of the symmetry operation $(S^{\perp}, \vec{s}^{\perp})$ in $U^{\perp}$.

<u>Definition 5.3:</u> The coset $\vec{p}^{\perp} + U$ is called the symmetry support of the symmetry operation $(S, \vec{s})$.

<u>Definition 5.4:</u> If the screw vector $\vec{t} = 0$ then the symmetry operation $(S, \vec{s})$ is called a rotation or, if $d = n-1$, a reflection. If $\vec{t} \neq 0$ then $(S, \vec{s})$ is called a screw rotation or, if $d = n-1$, a glide reflection.

We give in the following a practical method to calculate the symmetry support of a general symmetry operation

$(S,\vec{s})$. Let k be the order of the rotation S. The screw vector $\vec{t}$ is given by

$$\vec{t}:=S^{k-1}\vec{s}+\ldots+S\vec{s}+\vec{s}.$$

If $\vec{t}=0$ then $(S,\vec{s})$ has at least one fixed point. Otherwise if $\vec{t}\neq 0$ then $(S,\vec{s}-\vec{t}/k)$ has at least one fixed point. In order to calculate the fixed points we set

$$\vec{p}=(S,\vec{s}-\vec{t}/k)\vec{p}.$$

It follows that

$$(I-S)\vec{p}-\vec{s}+\vec{t}/k=0.$$

We set $A:=I-S$ and $\vec{v}:=-\vec{s}+\vec{t}/k$. Thus we have

$$A\vec{p}+\vec{v}=0.$$

We use the method of section 5.4 to solve this equation. The scheme looks as follows:

	μ_1	$\ldots$	μ_n	ζ
ξ_1	a_{11}	$\ldots$	a_{1n}	v_1
.	.		.	.
.	.		.	.
.	.		.	.
ξ_n	a_{n1}	$\ldots$	a_{nn}	v_n

Suppose that $\xi_1,\ldots,\xi_m$ can be exchanged with the algorithm of Gauss (cf. section 5.4). After m steps we obtain the following scheme where the columns 1 to m already have been dropped (possibly after renumbering).

	μ_{m+1}	$\ldots$	μ_n	ζ
μ_1	q_{11}	$\ldots$	q_{1d}	w_1
.	.		.	.
.	.		.	.
.	.		.	.
μ_m	q_{m1}	$\ldots$	q_{md}	w_m
ξ_{m+1}				
.				
.		0		0
.				
ξ_n				

The general solution is given by

$$\begin{pmatrix} \mu_1 \\ \cdot \\ \cdot \\ \cdot \\ \mu_m \end{pmatrix} = \begin{pmatrix} q_{11} \cdots q_{1d} \\ \cdot \qquad \cdot \\ \cdot \qquad \cdot \\ \cdot \qquad \cdot \\ q_{m1} \cdots q_{md} \end{pmatrix} \begin{pmatrix} \mu_{m+1} \\ \cdot \\ \cdot \\ \cdot \\ \mu_n \end{pmatrix} + \begin{pmatrix} w_1 \\ \cdot \\ \cdot \\ \cdot \\ w_m \end{pmatrix}$$

We can find d+1 fixed points f_i by choosing suitable real values for the $\mu_{m+1},\ldots,\mu_n$ such that the linear hull of the $\vec{f}_i-\vec{f}_1$, $i=2,\ldots,d+1$, span the subspace U. The symmetry support of $(S,\vec{s})$ then is the coset $\vec{f}_1+U$.

6. Crystallographic Point Groups

We now use the important concept of a mathematical group. The group elements are the symmetry operations. The group multiplication is the composition of symmetry operations as described in section 1.4.

6.1. Definitions

Let $X \subset E^n$ be a regular point system and let G be a space group which acts transitively on X. If P is a subgroup of G of finite order then the following theorem holds:

Theorem 6.1: The symmetry operations $(S,\vec{s})$ of a subgroup P of G of finite order have a common fixed point z.

Proof: Let h be the order of P. As h is finite P acts transitively only on a finite subset M of X. It follows that M is bounded and can be enclosed inside some ball of radius L. The centroid z of M is given by

$$z := \frac{1}{h} \sum_{x_i \in M} x_i .$$

The centroid z remains invariant under the action of P because every symmetry operation $(S,\vec{s}) \in P$ merely permutes the points $x_i \in M$,

$$(S,\vec{s})z = \frac{1}{h} \sum_{x_i \in M} (S,\vec{s})x_i = z.$$

Hence, the centroid z has the required property.

Clearly the fixed point z need not belong to the regular point system X. We can assume that the origin 0 is in the fixed point z. In reference to this origin, every symmetry operation of P must have the form (S,0) since otherwise the origin 0 would not remain fixed. In what follows we always assume the origin to be in the fixed point z and we only consider the rotation part S of the symmetry operation.

By definition 1.2, a group of symmetry operations which leaves at least one point fixed is called a point group.

Definition 6.1: A crystallographic point group in E^n is a group of symmetry operations which brings a translation lattice T^n into self-coincidence and leaves at least one point fixed.

By theorem 3.1, the crystallographic point groups are groups of finite order.

For each rotation S_i of a point group P we can determine the symmetry support U_i (cf. section 5.4). Several rotations $S_i \in P$ may have identical symmetry supports. We take the weighted union of all U_i,

$$M' := \bigcup_{S_i \in P} U_i .$$

Weighted means that for each subspace $U \subset M'$ a weight is assigned which counts the number of occurrences of identical symmetry supports. If the symmetry support U occurs k times then we say that U is a k-fold symmetry support.

Definition 6.2: The symmetry scaffolding of a point group P is the weigthed union of the symmetry supports of all symmetry operations $S \in P$.

In crystallography the term "symmetry element" is used for the symmetry scaffolding of a cyclic group. We avoid this term because of its confusion with the group elements.

We note that the symmetry support of a point group P is the intersection of the symmetry supports of all symmetry operations $S \in P$.

We represent the rotation S in E^n through a unimodular nxn matrix which carries the point x into y.

$$y = Sx.$$

In this way a point group P is represented through a matrix group. The coefficients of these matrices depend on the choice of the basis vectors. If $\vec{a}_1', \ldots, \vec{a}_n'$ is another choice with

$$\begin{pmatrix} \vec{a}_1' \\ \cdot \\ \cdot \\ \cdot \\ \vec{a}_n' \end{pmatrix} = \begin{pmatrix} a_{11} \ldots a_{1n} \\ \cdot \quad\quad \cdot \\ \cdot \quad\quad \cdot \\ \cdot \quad\quad \cdot \\ a_{n1} \ldots a_{nn} \end{pmatrix} \begin{pmatrix} \vec{a}_1 \\ \cdot \\ \cdot \\ \cdot \\ \vec{a}_n \end{pmatrix},$$

where A is a non-singular nxn matrix, then the coordinates of the point x' referred to the new basis are transformed with the contragredient matrix A^o,

$$x' = A^o x.$$

The rotation matrix S, referred to the new basis,is transformed according to (cf. section 3.2)

$$S' = A^o S A^t.$$

Because of physical reasons we require that det(A)=+1. The transformation A thus conserves the chirality of the basis. We say that S' and S are equivalent representations of one and the same symmetry operation. This transformation is applied to all symmetry operations $S \in P$. We say that A transforms P onto P' and we write

$$P' = A^o P A^t.$$

We have to emphasize that every symmetry operation maps E^n onto itself through its matrix representation. Therefore, equivalence of symmetry groups corresponds to the equivalence of their matrix representations. This distinguishes symmetry groups from abstract groups (cf. section 6.6).

<u>Definition 6.3:</u> Two point groups P and P' are equivalent and belong to the same geometric class if there exists a non-singular matrix W that transforms P onto P',

$$P' = WPW^{-1}.$$

Synonymous to geometric class $\mathbb{Q}$-class is used.

An important property of equivalent point groups is stated in the following theorem.

<u>Theorem 6.2:</u> If two point groups P and P' are equivalent then they are orthogonal equivalent,

$$P' = RPR^{-1}.$$

Proof: We use the fact that every non-singular matrix W can be written as a product W=RC, where R is an isometry and C is a positive-definte quadratic form (for a proof see Burc2). It follows that for every $S \in P$ we have

$$S' = RCSC^{-1}R^{-1}.$$

From the condition that $C=S^tCS$ (cf. section 1.3) it follows that $S^oC=CS$ hence,

$$S' = RS^oR^{-1}.$$

This is true for every $S \in P$ hence we have the required property.

$$P = RPR^{-1}.$$

<u>Theorem 6.3:</u> Equivalent point groups have congruent symmetry scaffoldings.

Proof: By theorem 6.2, there exists an isometry R such that $P' = RPR^{-1}$ holds. If U is the symmetry support of S^o then we have

$$S^oU = U = R^{-1}S'RU.$$

It follows that

$$S'RU=RU$$

hence, RU is the symmetry support of S'. Because R is an isometry the theorem 6.3 follows.

The above theorem 6.3 reveals that the symmetry scaffolding can be used to identify point groups.

A group of finite order is completely specified by its multiplication table. In general however, a few group elements are sufficient to generate the whole group.

<u>Definition 6.4:</u> A set M of symmetry operations is called a generating set of the point group P if every symmetry operation of P can be obtained as a product of symmetry operations of M.

Point groups may contain proper and improper symmetry operations (cf. section 1.3).

<u>Definition 6.5:</u> Point groups of the first kind are groups which contain only proper symmetry operations. Point groups of the second kind are groups which also contain improper symmetry operations.

Theorem 6.4: Every point group P of the second kind contains a point group P^+ of the first kind as a normal subgroup of index 2.

Proof: Every point group of the second kind contains proper rotations because the product of two improper rotations is a proper rotation. The proper rotations form a subgroup of the first kind. For any improper rotation $S^- \in P$ it holds that $S^-P=P$. Therefore, the numbers of proper and improper rotations must be equal. It follows that P^+ is of index 2 hence, it is normal in P.

6.2. Point groups in E^2

We use the Schönflies notation to designate the point groups. In 2-dimensional space E^2 the proper rotation S^+ generates the cyclic group C_k, where "C" stands for "cyclic" and the subscript k gives the order of S^+ (for point groups in E^2 we use the symbols of the 2+1 reducible point groups in E^3). The cyclic group C_k is a point group of the first kind.

Next we include the reflection and construct the point groups of the second kind. Let S^- represent the reflection in a mirror line m. The reflection S^- generates the cyclic group C_s of order $|C_s|=2$. The subscript "s" stands for the german word "Spiegel" (mirror). A point group of the first kind C_k may be combined with S^- to give a new group P of the second kind. The reflection S necessarily must leave the origin fixed but otherwise there are no restrictions in E^2. By theorem 6.4, the cyclic group C_k is a subgroup of P of index 2. The coset decomposition becomes

$$P = C_k \cup C_k S^-.$$

The symmetry operations of the coset C_kS^- are reflections. Each coset has the same number of symmetry operations hence, the order of the point group P is $|P|=2k$. Let $\varphi=360°/k$ be the angle of rotation of S^+. Then the symmetry support of S^+S^- is a mirror line m' rotated through the angle $\varphi/2$ out of the first mirror line m. The successive reflections in both mirror lines correspond to a proper rotation with rotation angle φ,

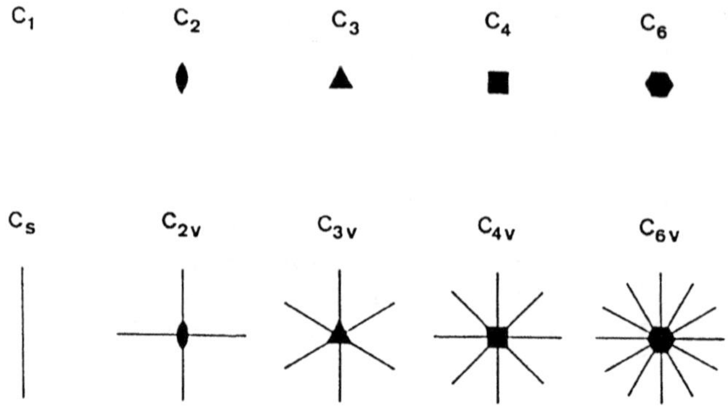

Figure 6.1. Representatives of the ten geometric crystal classes in the plane E^2 together with their graphic representation

$$(S^+S^-)S^- = S^+(S^-S^-) = S^+.$$

The point groups of the second kind are denoted by C_{kv}, where "v" means a "vertical" mirror line.

Among all point groups in E^2 we now determine the crystallographic point groups and we classify them according to the geometric crystal classes. By theorem 5.4, only rotations of orders 2,3,4, and 6 respectively are crystallographic rotations in E^2. It follows that there are ten non-equivalent geometric crystal classes in the plane E^2 which are shown in Figure 6.1.

6.3. Point groups in E^3

In three-dimensional space E^3 the proper rotations S^+ give rise to the cyclic groups C_k. A possible representation of a proper rotation S^+ is

$$S^+ := \begin{vmatrix} \cos 2\pi/k & -\sin 2\pi/k & 0 \\ \sin 2\pi/k & \cos 2\pi/k & 0 \\ 0 & 0 & 1 \end{vmatrix} .$$

The cyclic group generated by S^+ has order $|C_k|=k$. The symmetry scaffolding of C_k is called a k-fold rotation axis (cf. section 5.4).

In order to derive all point groups in E^3 we determine first the point groups of the first kind. Two cyclic groups C_k and C_l may be combined to give a new point group P^+ of the first kind. Let Q^+ and R^+ be the generating rotations of C_k and C_l respectively and let A, B be the corresponding rotation axes. By theorem 6.1, the point group P^+ has at least one fixed point z which we assume to be the origin. It follows that the rotation axes A and B must have the origin O in common. If A is not parallel to B then A intersects B in O and P^+ has exactly one fixed point. The possible combinations of intersecting rotation axes are obtained through the following construction which is due to Euler.

By the group property we have that $Q^+R^+=S^+$. The resulting rotation S^+ has order m and rotation axis C which necessarily must go through the origin O. The action of the three rotations on a point x can be obtained through successive reflections in three mirror planes each one containing two of the rotation axes as can be seen from Figure 6.2.

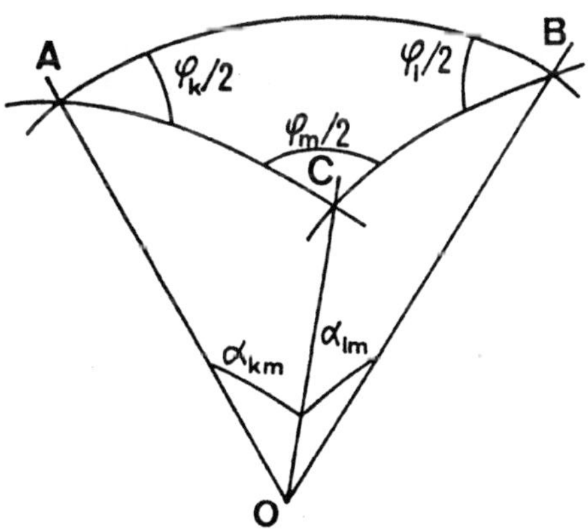

Figure 6.2. Euler's construction to obtain the possible combinations of rotation axes

The three mirror planes form a spherical triangle with the rotation axes as edges. The angles between every pair of intersecting mirror planes are $\varphi_k/2$, $\varphi_l/2$, and $\varphi_m/2$

respectively. Using the spherical cosine formula the angles between the rotation axes are obtained

$$\cos\alpha_{kl} = \frac{\cos\varphi_m/2 + \cos\varphi_k/2 \, \cos\varphi_l/2}{\sin\varphi_k/2 \, \sin\varphi_l/2} \, .$$

The result of these calculations are summerized in Table 6.1.

Table 6.1. The possible solutions of the spherical cosine formula

k	l	m	$\varphi_k/2$	$\varphi_l/2$	$\varphi_m/2$	α_{lm}	α_{km}	α_{kl}
2	2	2	90°	90°	90°	90°	90°	90°
2	2	3	90°	90°	60°	90°	90°	60°
2	2	4	90°	90°	45°	90°	90°	45°
2	2	.	90°	90°	.	90°	90°	.
2	2	.	90°	90°	.	90°	90°	.
2	2	.	90°	90°	.	90°	90°	.
2	2	n	90°	90°	180°/n	90°	90°	180°/n
2	3	3	90°	60°	60°	70°32'	54°44'	54°44'
2	3	4	90°	60°	45°	54°44'	45°	35°16'
2	3	5	90°	60°	36°	37°22'	31°43'	20°54'

The various types of point groups of the first kind as they result from Table 6.1 are illustrated in Figures 6.3, a)-d). They are:

a) The dihedral group D_k of all proper rotations of a regular k-faced prism.

b) The tetrahedral group T of all proper rotations of the regular tetrahedron.

c) The octahedral group O of all proper rotations of the regular octahedron.

d) The icosahedral group I of all proper rotations of the regular icosahedron.

e) For non-integral values of k, l, or m the group K of all proper rotations of the sphere results. This group has infinite order.

The tetrahedron, octahedron, and icosahedron are three among the five Platonic solids. Two of them have a dual

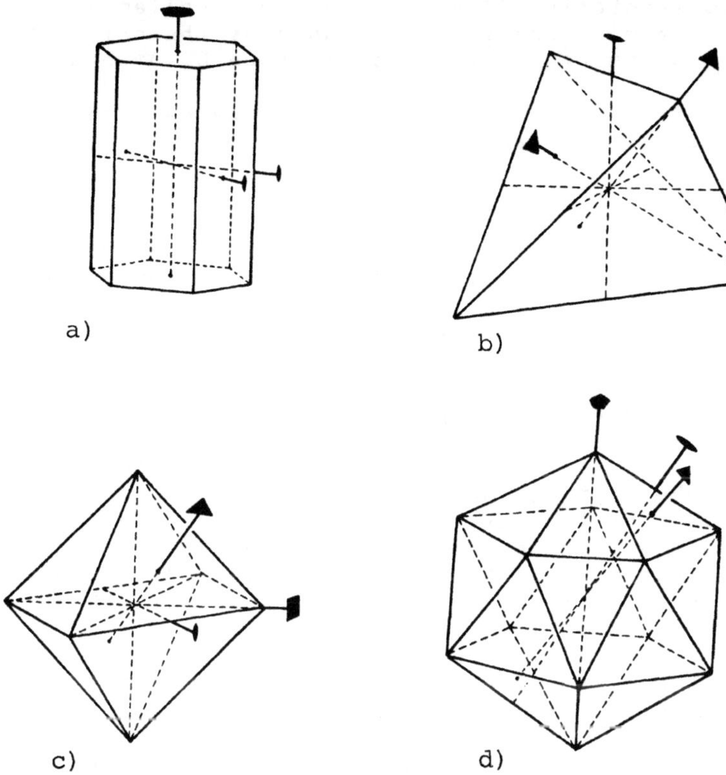

Figure 6.3. The point groups of the first kind. a) k-faced prism, b) tetrahedron, c) octahedron, d) icosahedron

counterpart. The octahedron is dual to the cube hence, its group of proper rotations is also O. The icosahedron is dual to the dodecahedron.

We now determine the point groups of the second kind. By theorem 6.4, a point group P of the second kind has a point group P^+ of the first kind as a normal subgroup of index 2. The coset decomposition becomes:

$$P = P^+ \cup P^+S^-.$$

As a representative S^- we choose the reflection in a mirror plane m. Necessarily m contains the origin O. The reflection S^- generates the cyclic group C_s of order 2.

Because P^+ is a normal subgroup of P we have to require that S^- leaves P^+ invariant,

$$S^-P^+(S^-)^{-1}=P^+.$$

Geometrically this means that S^- maps the symmetry scaffolding of P^+ onto itself. It follows that there are only a few non-equivalent possibilities to add a mirror plane m to the symmetry scaffolding of P^+.

We take the cyclic group C_k. The k-fold rotation axis is assumed to be vertical. There exist two possible settings of m. Either it can be perpendicular two the k-fold rotation axis. In this case it is called a "horizontal" mirror plane m_h. The point group C_{kh} results. Or it can contain the k-fold rotation axis. It is then called a "vertical" mirror plane m_v. The point group C_{kv} results. If S^+ is the generating rotation of C_k then the symmetry support of S^+S^- is the mirror plane m'_v which is rotated through the angle 180°/k out of the mirror plane m_v.

We take the dihedral group D_k. The k-fold rotation axis is assumed to be vertical. Either we can add a horizontal mirror plane m_h. The point group D_{kh} results. Because the twofold rotation axes are prependicular to the k-fold rotation axis also vertical mirror planes m_v are generated which intersect m_h along the twofold axes. Or we can add a mirror plane m_d which contains the k-fold rotation axis and which bisects the angle between two adjacent twofold rotation axes. Thus the mirror plane m_d is "diagonal" with respect to the twofold rotation axes. The point group D_{kd} results.

The tetrahedral group T is set up with a twofold rotation axis vertical. Either we can add a horizontal mirror plane m_h perpendicular to the twofold axis. The point group T_h results. Or we can add a mirror plane m_d which contains a threefold axis and which is diagonal to two adjacent twofold axes. The point group T_d results.

The octahedral group O is set up with a fourfold axis vertical. We can add a horizontal mirror plane m_h perpendicular to the fourfold axis. The point group O_h results. The other possibilities are equivalent and do not give new groups.

The icosahedral group I is set up with a twofold axis vertical. We can add a horizontal mirror plane m_h perpendicular to the twofold axis. The point group I_h results. There is no other possibility.

To the group K of all proper rotations of a sphere we can add a mirror plane of arbitrary orientation which we call m_h. The point group K_h results.

Finally we have to consider the general rotoreflection S^- (cf. section 5.4). A possible representation is:

$$S^- := \begin{pmatrix} \cos 2\pi/k & -\sin 2\pi/k & 0 \\ \sin 2\pi/k & \cos 2\pi/k & 0 \\ 0 & 0 & -1 \end{pmatrix}.$$

The rotoreflection S^- generates a cyclic point group S_k.

For k=2 the rotoreflection S^- is the central inversion $-I$. It generates the cyclic group C_i of order 2 representing a center of symmetry.

For $k=4m+2$, $m \in \mathbb{N}$, the cyclic group S_k can be generated through the cyclic group C_l of order $l=k/2$ and the central inversion $-I$. Thus $S_k=C_{li}$, if $k=4m+2$.

If k is odd then S_k can be generated through the cyclic group C_k and a reflection in a horizontal mirror plane m_h. Thus $S_k=C_{kh}$, if k is odd.

Only if $k=4m$ then S_k cannot be generated by other cyclic groups.

Table 6.2. Flow chart to determine the point groups in E^3

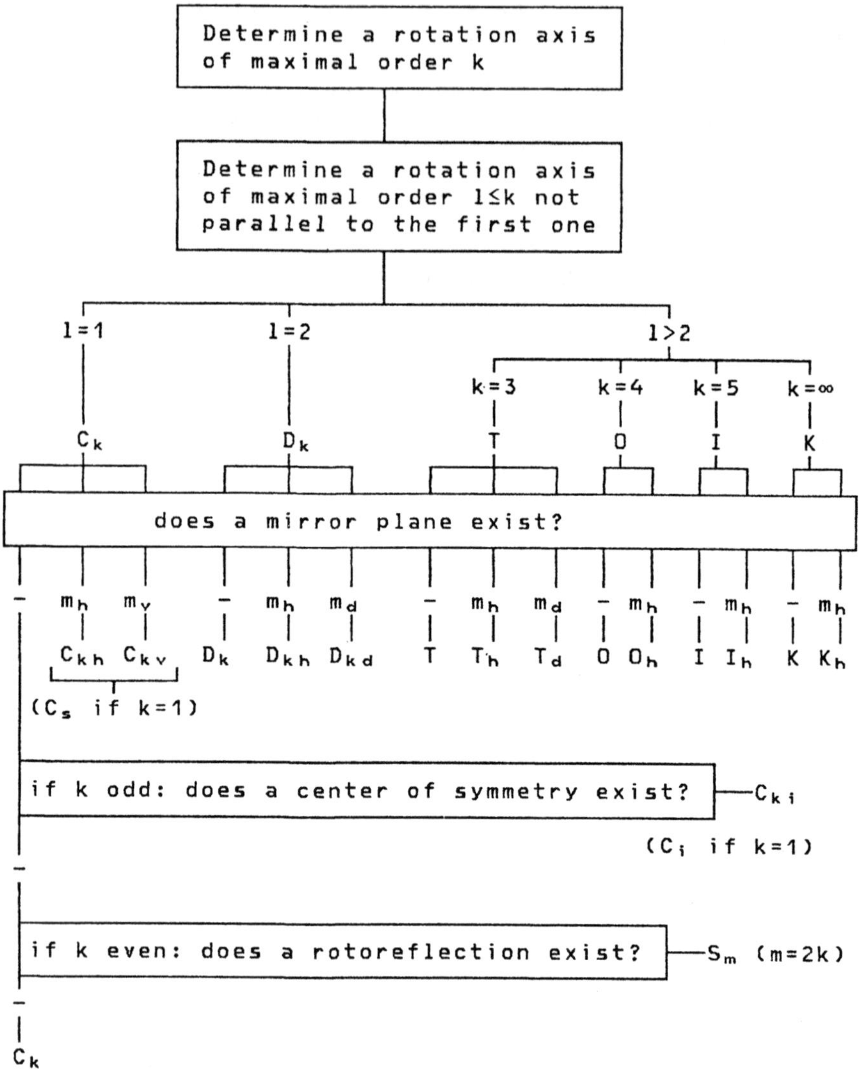

Remark: The k-fold rotation axis is assumed to be vertical. Only in the T and I groups the twofold rotation axis is taken to be vertical. If both horizontal and vertical mirror planes occur the horizontal mirror plane is taken.

There exists an infinite number of point groups in E^3 yet, these may be grouped into a few series as is shown in the flow chart of Table 6.2 which helps to determine the point groups in E^3. This flow chart uses the concept of generating symmetry operations and therefore, only a few steps are necessary to completely determine a point group in E^3.

Among all point groups in E^3 we now determine the crystallographic point groups and we classify them according to the geometric crystal classes. In E^3 we have the same orders of rotations as in E^2, viz.: 2,3,4, and 6 respectively. This criterion is sufficient to characterize the crystallographic point groups in E^3.

Among the cyclic point groups we find the following geometric crystal classes:

C_1, C_2, C_3, C_4, C_6, C_s, C_i, C_{3h}, S_4, C_{3i},

and, if we add a mirror plane, we get in addition

C_{2h}, C_{4h}, C_{6h}, C_{2v}, C_{3v}, C_{4v}, C_{6v}.

Among the dihedral point groups we have the following geometric crystal classes:

D_2, D_3, D_4, D_6, D_{2h}, D_{3h}, D_{4h}, D_{6h}, D_{2d}, D_{3d}.

We note that D_{4d} and D_{6d} have rotoreflections of order 8 and 12 respectively therefore, they are not crystallographic.

The tetrahedral and octahedral point groups belong to the geometric crystal classes:

T, T_h, T_d, O, O_h.

Altogether there exist 32 geometric crystal classes in E^3. The relation between these crystal classes can be seen from the group-subgroup diagram in Figure 6.4.

Alternatively to the Schönflies notation for point groups also the international notation of Hermann and Mauguin is used (for details the reader is referred to the International Tables, Vol. A, Hahn1). Both kinds of symbols have in common that they specify for each geometric crystal class a set of generating symmetry operations.

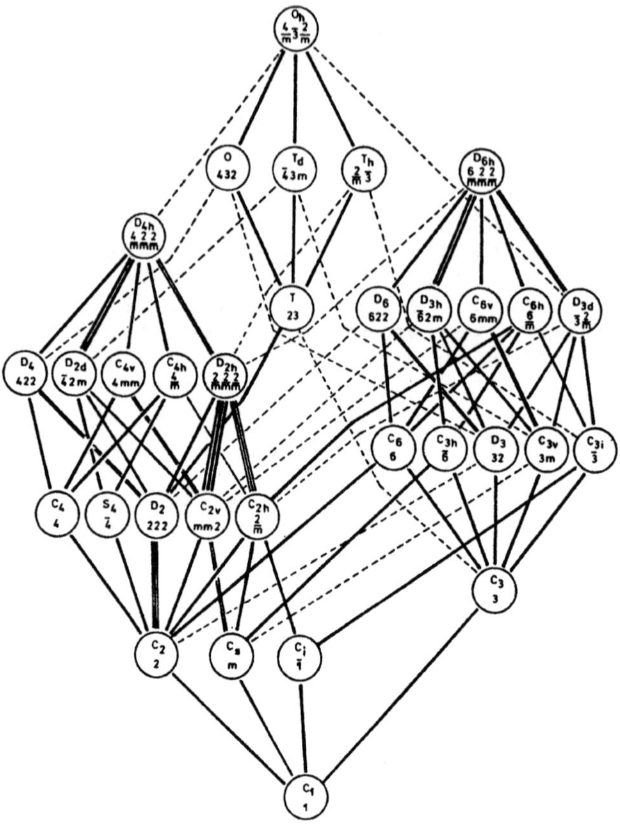

Figure 6.4. Diagram of the group-subgroup relations between the geometric crystal classes in E^3

6.4. Point groups in E^n

In spaces of higher dimensions the point groups may not as easily be visualized as in two- and three-dimensional space. In what follows we present a geometric derivation of the point groups in E^n.

By theorem 5.3, every proper rotation S^+ in E^n can be represented through rotations in a set of mutually perpendicular principal planes $\pi_1, \ldots, \pi_m$ with rotations $\varphi_1, \ldots, \varphi_m$ respectively and one-dimensional invariant subspaces. Referred to a cartesian coordinate system

$\vec{e}_1,\ldots,\vec{e}_n$ within these subspaces we obtain the following representation of S^+ (cf. section 5.3),

$$S^+ := \begin{pmatrix} \cos\varphi_1 & -\sin\varphi_1 & & & & \\ \sin\varphi_1 & \cos\varphi_1 & & 0 & & \\ & & \cdot & & & \\ & 0 & & \cdot & & \\ & & & \cos\varphi_m & -\sin\varphi_m & \\ & & & \sin\varphi_m & \cos\varphi_m & \\ & & & & & I_h \end{pmatrix}.$$

In section 6.2 we had replaced S^+ through the successive reflection in two mirror lines m and m'. In an analogous way we now replace φ_1 by two successive involutions Z_1 and Z_2 which we construct in the following way (an involution is a symmetry operation of order 2). We choose in π_1 a vector of unit length which we call the root vector $\vec{r}_1$. The involution Z_1 is determined in such a way that it leaves $\vec{r}_1$ fixed and inverts the orthogonal complement $U^\perp$ of $\vec{r}_1$ in E^n. We are free to choose $\vec{r}_1=\vec{e}_1$. The involution Z_1 is then represented through a diagonal matrix of the following form,

$$Z_1 := \begin{pmatrix} 1 & & & & \\ & -1 & 0 & & \\ & & \cdot & & \\ & 0 & & \cdot & \\ & & & & -1 \end{pmatrix}.$$

For the involution Z_2 we determine the root vector $\vec{r}_2\in\pi_1$ in such a way that the angle $\sphericalangle(\vec{r}_1,\vec{r}_2)$ is $\varphi_1/2$. By conjugation of Z_1 with

$$V := \begin{pmatrix} \cos\varphi_1/2 & \sin\varphi_1/2 & & & \\ -\sin\varphi_1/2 & \cos\varphi_1/2 & & 0 & \\ & & 1 & & \\ & 0 & & \cdot & \\ & & & & \cdot \\ & & & & & 1 \end{pmatrix},$$

we obtain $Z_2 := VZ_1V^{-1}$,

$$Z_2 = \begin{pmatrix} \cos\varphi_1 & -\sin\varphi_1 & & & \\ -\sin\varphi_1 & -\cos\varphi_1 & & 0 & \\ & & -1 & & \\ & 0 & & \cdot & \\ & & & & \cdot \\ & & & & & -1 \end{pmatrix}.$$

It follows that

$$Z_1 Z_2 = \begin{pmatrix} \cos\varphi_1 & -\sin\varphi_1 & & & \\ \sin\varphi_1 & \cos\varphi_1 & & 0 & \\ & & 1 & & \\ & 0 & & \cdot & \\ & & & & \cdot \\ & & & & 1 \end{pmatrix} .$$

In a similar way the remaining rotations $\varphi_2, \ldots, \varphi_m$ are replaced by involutions. As a result we have that S^+ can be written as a product of 2m involutions,

$$S^+ = Z_1 \ldots Z_{2m} .$$

By the group property the product of two proper rotations Q^+, R^+ is again a proper rotation,

$$Q^+ R^+ = S^+ .$$

Let Q^+, R^+ be generating rotations of a point group P^+ of finite order. We denote by π_{1i}, π_{2i}, π_{3i}, $i=1,\ldots,h\leq m$ the principal planes of Q^+, R^+, S^+, and by φ_{1i}, φ_{2i}, φ_{3i} the corresponding rotation angles. If Q^+ and R^+ have identical principal planes then so does S^+. We now assume that Q^+ and R^+ have different principal planes. Two cases have to be considered.

1. The principal planes intersect in straight lines. Assume that π_{11} and π_{21} intersect in a straight line. We are free to choose $\pi_{11} \cap \pi_{21} = \lambda \vec{r}_1$. Thus the root vector $\vec{r}_1$ is common to both principal planes. The other two root vectors $\vec{r}_2 \in \pi_{11}$ and $\vec{r}_3 \in \pi_{21}$ are now determined by the angles $\varphi_{11}/2$ and $\varphi_{21}/2$ respectively. Let Z_1, Z_2, and Z_3 be the corresponding involutions. From $Q^+ R^+ = S^+$ it follows that

$$Z_2 Z_1 \; Z_1 Z_3 = Z_2 Z_3 ,$$

hence, $\vec{r}_2$ and $\vec{r}_3$ determine the principal plane π_{31}. We have that $\pi_{11} \cap \pi_{31} = \lambda \vec{r}_2$ and $\pi_{21} \cap \pi_{31} = \lambda \vec{r}_3$. The carrier space of π_{11}, π_{21}, and π_{31} is a three-dimensional subspace $U \subset E^n$. Let S^{n-1} be the unit sphere in E^n (it is the (n-1)-dimensional surface of the unit ball in E^n). It follows that $U \cap S^{n-1} = S^2$. Thus the involutions Z_1, Z_2, and Z_3 leave a sphere S^2 invariant and therefore, they have to belong either to the dihedral (D), tetrahedral (T), octahedral (O), or icosahedral (I) point group. It follows that the angles $\sphericalangle(\vec{r}_i, \vec{r}_j)$, $i,j=1,2,3$, $i<j$, must have well defined values.

We now look at the second set of principal planes π_{12}, π_{22}, π_{32} which are orthogonal to π_{11}, π_{21}, and π_{31} respectively. Let $U^{\perp}$ be the orthogonal complement of U. There exists a unique vector $\vec{p}\in U$ orthogonal to π_{12}. It follows that π_{12} intersects $U^{\perp}$ at least in a one-dimensional subspace. The same holds for π_{22} and π_{32}. We have $\pi_{12}\cap\pi_{22}=\lambda\vec{r}_4$, $\pi_{12}\cap\pi_{32}=\lambda\vec{r}_5$, and $\pi_{22}\cap\pi_{32}=\lambda\vec{r}_6$. At least one of these principal planes has to be a subspace of $U^{\perp}$ because otherwise at least two of the root vectors $\vec{r}_4$, $\vec{r}_5$, or $\vec{r}_6$ are not contained in $U^{\perp}$ and therefore, one of these principal planes does not intersect $U^{\perp}$ in a one-dimensional subspace which is a contradiction. It follows that the carrier space of $\vec{r}_1,\ldots,\vec{r}_6$ is at least a five dimensional subspace of E^n. In a similar way we can account for the remaining principal planes. The distribution of the root vectors $\vec{r}_i$ on the sphere S^{n-1} gives the necessary restrictions for possible point groups of finite order. We have to require that always k of the corresponding involutions Z_i leave a sphere S^{k-1} invariant.

2. The principal planes intersect only in the origin. This can occur if two or more principal planes π_{1i} of a rotation Q^+ have a congruent system of root vectors. In this case there exists an involution Y which interchanges these principal planes. Assume Y interchanges π_{11} with π_{12} then, referred to a cartesian coordinate system within these principal planes, the involution Y may have the following form

$$Y:=\begin{pmatrix} 0 & 0 & 0 & 1 & \\ 0 & 0 & 1 & 0 & \\ 0 & 1 & 0 & 0 & 0 \\ 1 & 0 & 0 & 0 & \\ & & 0 & & I_h \end{pmatrix}.$$

Among all point groups P in E^n we have to determine the crystallographic point groups. The following two criteria can be used to decide if a point group is crystallographic:

1. Every rotation $S\in P$ has to be crystallographic (cf. section 5.3).

2. A crystallographic point group cannot contain involutions Z_i which are in principal planes of rotation orders h=5, or h>6.

Example 6.1: We now give some examples of point groups in four-dimensional space E^4. Assume that the principal planes π_{11}, π_{21}, and π_{31} intersect, $\pi_{11}\cap\pi_{21}=\lambda\vec{r}_1$, $\pi_{11}\cap\pi_{31}=\lambda\vec{r}_2$ and $\pi_{21}\cap\pi_{31}=\lambda\vec{r}_3$. The root vectors $\vec{r}_1$, $\vec{r}_2$ and $\vec{r}_3$ span a three-dimensional carrier space $U\subset E^4$. The corresponding involutions are Z_1, Z_2 and Z_3 respectively. Let $\lambda\vec{p}$ be the orthogonal complement of U in E^4. It follows that in E^4 at most one of the rotations Q^+, R^+ and S^+ can have a second principal plane because suppose two of them have a second principal plane then these necessarily intersect in $\lambda\vec{p}$. Hence, a further principal plane would exist which does not contain $\vec{p}$ which is a contradiction. Assume that the additional principal plane is π_{12}. We have two further root vectors $\vec{r}_4$ and $\vec{r}_5$ in π_{12} with involutions Z_4 and Z_5. The root vectors $\vec{r}_1,\ldots,\vec{r}_5$ determine two adjacent spherical tetrahedra on the sphere S^3. We now require that always three of these involutions $Z_1,\ldots,Z_5$ leave a sphere S^2 invariant. Therefore, the angles between these root vectors must correspond to the angles in the point groups D, T, O, or I. Always four among these involutions leave a sphere S^3 invariant. Thus they are generators of a point group in E^4. Possible solutions are shown in Figures 6.5, a)-d) which show the distribution of the root vectors on the sphere S^3. The intersection of the root vectors $\vec{r}_i$, $i=1,\ldots,5$, with S^3 are identified with the points r_i. These points are the vertices of two adjacent spherical tetrahedra. The lines represent great circles, viz.: $C_{11}=\pi_{11}\cap S^3$ and $C_{12}=\pi_{12}\cap S^3$. We refer the root vectors to a cartesian coordinate system $\vec{e}_1,\ldots,\vec{e}_4$. For a given root vector $\vec{r}:=[\rho_1,\rho_2,\rho_3,\rho_4]$ the involution Z is obtained as follows: We determine a new basis $\vec{e}'_1,\ldots,\vec{e}'_4$ with the transformation A where $\vec{e}'_1:=\vec{r}$ and $\vec{e}'_2$, $\vec{e}'_3$, $\vec{e}'_4$ are orthogonal to $\vec{r}$. In reference to the new coordinate system, the involution Z' has the form

$$Z':=\begin{vmatrix}1&0&0&0\\0&-1&0&0\\0&0&-1&0\\0&0&0&-1\end{vmatrix}.$$

Referred to the original basis the involution Z becomes (cf. section 3.2)

$$Z=A^tZ'A^o.$$

In Figure 6.5. a) we have the root vectors

$$\vec{r}_1:=\sqrt{2}[1,-1,0,0] \qquad \vec{r}_2:=\sqrt{2}[1,0,1,0]$$

$\vec{r}_3 := \sqrt{2}[1,1,0,0]$ $\vec{r}_4 := \sqrt{2}[0,0,1,1]$

$\vec{r}_5 := \sqrt{2}[0,1,0,1]$.

The angles between the root vectors are shown in the Figure. (not indicated are the angles $\sphericalangle(\vec{r}_1,\vec{r}_5)=90°$ and $\sphericalangle(\vec{r}_2,\vec{r}_4)=60°$). These angles correspond to D (dihedral) and T (tetrahedral) point groups. We calculate the corresponding involutions

$$Z_1 := \begin{pmatrix} 0 & -1 & 0 & 0 \\ -1 & 0 & 0 & 0 \\ 0 & 0 & -1 & 0 \\ 0 & 0 & 0 & -1 \end{pmatrix} \qquad Z_2 := \begin{pmatrix} 0 & 0 & 1 & 0 \\ 0 & -1 & 0 & 0 \\ 1 & 0 & 0 & 0 \\ 0 & 0 & 0 & -1 \end{pmatrix}$$

$$Z_3 := \begin{pmatrix} 0 & 1 & 0 & 0 \\ 1 & 0 & 0 & 0 \\ 0 & 0 & -1 & 0 \\ 0 & 0 & 0 & -1 \end{pmatrix} \qquad Z_4 := \begin{pmatrix} -1 & 0 & 0 & 0 \\ 0 & -1 & 0 & 0 \\ 0 & 0 & 0 & 1 \\ 0 & 0 & 1 & 0 \end{pmatrix}$$

$$Z_5 := \begin{pmatrix} -1 & 0 & 0 & 0 \\ 0 & 0 & 0 & 1 \\ 0 & 0 & -1 & 0 \\ 0 & 1 & 0 & 0 \end{pmatrix}.$$

The point group generated from the involutions Z_1, Z_2, Z_3, and Z_5, is the group B_4 of order 192 (The nomenclature is taken fom Coxeter (Coxe1)).

In Figure 6.5. b) we have the root vectors

$\vec{r}_1 := [1,0,0,0]$ $\vec{r}_2 := \sqrt{2}[1,1,0,0]$

$\vec{r}_3 := \sqrt{2}[0,1,1,0]$ $\vec{r}_4 := -[0,0,0,1]$

$\vec{r}_5 := \sqrt{2}[0,0,1,1]$.

The corresponding involutions are

$$Z_1 := \begin{pmatrix} 1 & 0 & 0 & 0 \\ 0 & -1 & 0 & 0 \\ 0 & 0 & -1 & 0 \\ 0 & 0 & 0 & -1 \end{pmatrix} \qquad Z_2 := \begin{pmatrix} 0 & 1 & 0 & 0 \\ 1 & 0 & 0 & 0 \\ 0 & 0 & -1 & 0 \\ 0 & 0 & 0 & -1 \end{pmatrix}$$

$$Z_3 := \begin{pmatrix} -1 & 0 & 0 & 0 \\ 0 & 0 & 1 & 0 \\ 0 & 1 & 0 & 0 \\ 0 & 0 & 0 & -1 \end{pmatrix} \qquad Z_4 := \begin{pmatrix} -1 & 0 & 0 & 0 \\ 0 & -1 & 0 & 0 \\ 0 & 0 & -1 & 0 \\ 0 & 0 & 0 & 1 \end{pmatrix}$$

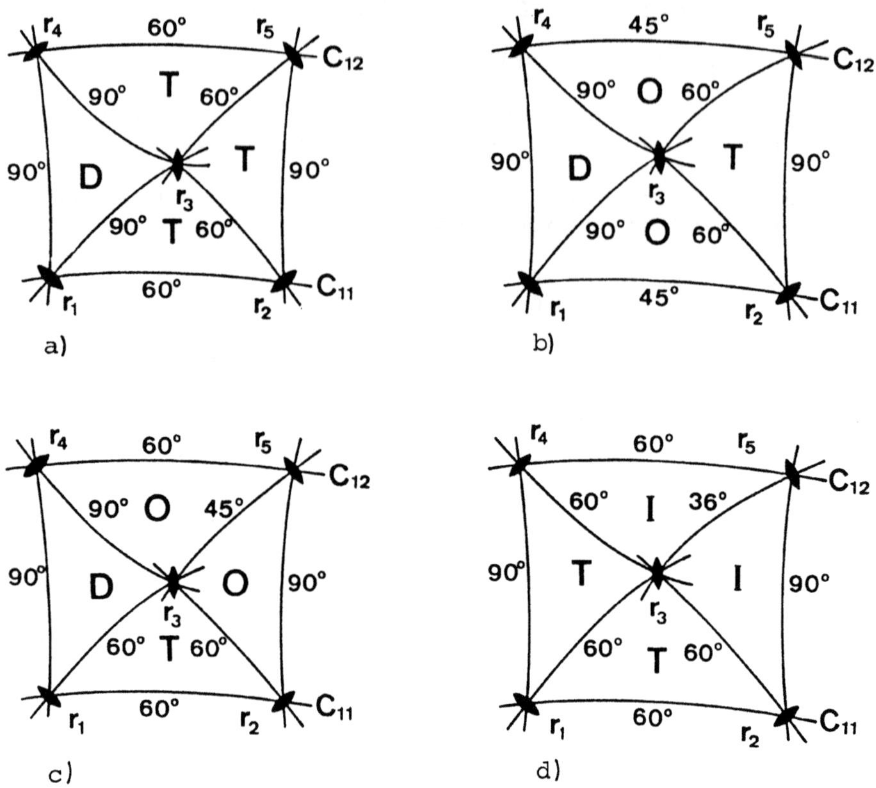

Figure 6.5. Generating involutions of some four-dimensional point groups

$$Z_5 := \begin{vmatrix} -1 & 0 & 0 & 0 \\ 0 & -1 & 0 & 0 \\ 0 & 0 & 0 & 1 \\ 0 & 0 & 1 & 0 \end{vmatrix} .$$

The point group generated from the involutions Z_1, Z_2, Z_3, and Z_5 is the group C_4 of order 384. It is the point group of the regular hypercube {3,3,4} which is dual to the crosspolytope {4,3,3}.

In Figure 6.5. c) we have the root vectors

$$\vec{r}_1 := \sqrt{2}[1,0,1,0] \qquad \vec{r}_2 := \sqrt{2}[1,1,0,0]$$

$\vec{r}_3 := \sqrt{2}[0,1,1,0]$ $\vec{r}_4 := [0,0,0,1]$

$\vec{r}_5 := [-\frac{1}{2},\frac{1}{2},\frac{1}{2},\frac{1}{2}]$.

The corresponding involutions are

$$Z_1 := \begin{pmatrix} 0 & 0 & 1 & 0 \\ 0 & -1 & 0 & 0 \\ 1 & 0 & 0 & 0 \\ 0 & 0 & 0 & -1 \end{pmatrix} \qquad Z_2 := \begin{pmatrix} 0 & 1 & 0 & 0 \\ 1 & 0 & 0 & 0 \\ 0 & 0 & -1 & 0 \\ 0 & 0 & 0 & -1 \end{pmatrix},$$

$$Z_3 := \begin{pmatrix} -1 & 0 & 0 & 0 \\ 0 & 0 & 1 & 0 \\ 0 & 1 & 0 & 0 \\ 0 & 0 & 0 & -1 \end{pmatrix} \qquad Z_4 := \begin{pmatrix} -1 & 0 & 0 & 0 \\ 0 & -1 & 0 & 0 \\ 0 & 0 & -1 & 0 \\ 0 & 0 & 0 & 1 \end{pmatrix}$$

$$Z_5 := \begin{pmatrix} -\frac{1}{2} & -\frac{1}{2} & -\frac{1}{2} & -\frac{1}{2} \\ -\frac{1}{2} & -\frac{1}{2} & \frac{1}{2} & \frac{1}{2} \\ -\frac{1}{2} & \frac{1}{2} & -\frac{1}{2} & \frac{1}{2} \\ -\frac{1}{2} & \frac{1}{2} & \frac{1}{2} & -\frac{1}{2} \end{pmatrix}.$$

The point group generated from the involutions Z_1, Z_2, Z_3, and Z_5 is the group C_4. The point group generated from the involutions Z_2, Z_3, Z_4, and Z_5 is the group F_4 of order 1152. It is the group of the regular 24-cell {3,4,3}. It has no dual counterpart.

In Figure 6.5. d) we have the root vectors

$\vec{r}_1 := \sqrt{2}[1,0,1,0]$ $\vec{r}_2 := \sqrt{2}[1,1,0,0]$

$\vec{r}_3 := \sqrt{2}[0,1,1,0]$ $\vec{r}_4 := [-a,a,a,b]$

$\vec{r}_5 := [-c,c,c,-d]$,

with $a:\sim 0.3535534$, $b:\sim 0.7905694$, $c:\sim 0.5720614$, and $d:\sim 0.1350454$.

The corresponding involutions are

$$Z_1 := \begin{pmatrix} 0 & 0 & 1 & 0 \\ 0 & -1 & 0 & 0 \\ 1 & 0 & 0 & 0 \\ 0 & 0 & 0 & -1 \end{pmatrix} \qquad Z_2 := \begin{pmatrix} 0 & 1 & 0 & 0 \\ 1 & 0 & 0 & 0 \\ 0 & 0 & -1 & 0 \\ 0 & 0 & 0 & -1 \end{pmatrix}$$

$$Z_3 := \begin{pmatrix} -1 & 0 & 0 & 0 \\ 0 & 0 & 1 & 0 \\ 0 & 1 & 0 & 0 \\ 0 & 0 & 0 & -1 \end{pmatrix} \qquad Z_4 := \begin{pmatrix} -\frac{3}{4} & -\frac{1}{4} & -\frac{1}{4} & -h \\ -\frac{1}{4} & -\frac{3}{4} & \frac{1}{4} & h \\ -\frac{1}{4} & \frac{1}{4} & -\frac{3}{4} & h \\ -h & h & h & \frac{1}{4} \end{pmatrix}$$

$$Z_5 := \begin{pmatrix} -i & -j & -j & k \\ -j & -i & j & -k \\ -j & j & -i & -k \\ k & -k & -k & -l \end{pmatrix},$$

with $h:\sim 0.559017$, $i:\sim 0.345492$, $j:\sim 0.654508$, $k:\sim 0.154508$, $l:\sim 0.963525$. The point group generated from the involutions Z_1, Z_2, Z_3, and Z_4 is the group A_4 of order 120. It is the group of the regular simplex {3,3,3}. We denote by A_4^* the group of order 240 generated by A_4 and the central inversion $-I$. The point group generated from the involutions Z_1, Z_2, Z_3, and Z_5 is the group G_4 of order 14400. It is the point group of the regular 600-cell {3,3,5} which is dual to the 120-cell {5,3,3}.

We observe that in E^4 a combination of the octahedral and the icosahedral point groups is not possible. It means that there exists no combination of spherical octahedra and icosahedra which tessellate the sphere S^3.

Remarkably in spaces of higher dimensions $n>4$ only three regular polytopes occur, viz.: the regular simplex with point group A_n of order $(n+1)!$ and the regular crosspolytope and its dual counterpart the regular hypercube with point group C_n of order $2^n n!$.

The point groups derived in case 1 can also be generated by reflections in a set of hyperplanes. The relation to the involutions Z_i is given by the property that $-IZ_i$ is a reflection in a hyperplane perpendicular to the root vector $\vec{r}_i$ (for details on groups generated by reflections the reader is referred to Coxeter (Coxe1)).

We note that in spaces of even dimension the central inversion $-I$ is a proper and Z_i is an improper involution. In spaces of odd dimension $-I$ is an improper and Z_i is a proper involution.

Similarly, as it was done in three-dimensional space, we can group the four-dimensional point groups in a finite number of series. A possible way to do this will be shown in the next section when discussing the root classes.

Among all point groups in E^4 we have to determine the crystallographic point groups and to classify them into the geometric crystal classes. The actual derivation of the crystal classes in E^4 however, started from the maximal irreducible integral matrix groups. In E^2 the crystal classes C_{4v} and C_{6v} are maximal. In E^3 the crystal class O_h is maximal. To O_h belong three non-equivalent integral matrix groups. There are three irreducible maximal crystal classes in E^4 having orders 1152, 288, and 240. Each of these gives rise to two non-equivalent integral matrix groups. The latter were determined by Dade (Dade1). Plesken and Pohst determined the maximal irreducible integral matrix groups up till n=9 (Ples1). The results of these calculations are given in Table 6.3.

Table 6.3. The numbers N of maximal irreducible integral matrix groups for each dimension n up to 9

n	1	2	3	4	5	6	7	8	9
N	1	2	3	6	7	17	7	26	20

The subgroups of the maximal integral matrix groups yield all geometric crystall classes. In four-dimensional space they were calculated by Bülow (Bülo1) using group theoretical computer programs.

The results of these calculations are summerized in the following theorem.

Theorem 6.5: There exist 2 in one-dimensional, 10 in two-dimensional, 32 in three-dimensional, and 227 geometric crystal classes in four-dimensional space.

6.5. Root classes

The root vectors $\vec{r}_i$ were introduced in section 6.4. as unit vectors in the principal planes. If two principal planes intersect along a one-dimensional subspace then we can choose a root vector in this subspace. However, for a given point group P, not all intersections of principal planes determine root vectors. We further require that the corresponding involutions Z leave the point group P invariant that is,

$$ZPZ^{-1} = P.$$

Example 6.2: In E^3 we consider the icosahedral group I. The root vectors lie in the twofold rotation axes which are given by the intersections of the principal planes corresponding to the rotations of order three and five. However, the intersections of the principal planes corresponding to different rotations of order three determine no root vectors as is shown in Figure 6.6.

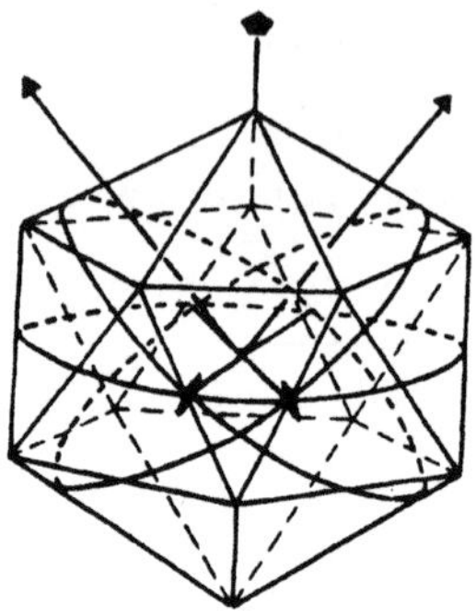

Figure 6.6. Root vectors in the icosahedral group I

Besides the involutions Z each of which leaves a one-dimensional subspace fixed there can exist other involutions in a point group. For example the central inversion -I is such an involution. In section 6.4 we introduced the involution Y which interchanges different principal planes. Another possible involution, we call it J, has the following form

$$J := \begin{pmatrix} I_h & 0 \\ 0 & -I_k \end{pmatrix} ,$$

where $h,k>1$ (if $h=1$ then J is an involution Z, if $k=1$ then J is a reflection in a hyperplane and -IJ is an involution Z). We do not write these involutions as products of additional involutions Z_i because, in this case, the Z_i are not uniquely defined. We only consider this kind of involutions if they are essential generators of the point group.

To every point group P of finite order we can assign a root system R which contains all root vectors,

$$R := \{\vec{r}_1, \dots, \vec{r}_s\}.$$

Thus we obtain a natural classification of the point groups.

Definition 6.6: Two point groups belong to the same root class if they have congruent root systems. Point groups having in addition essential involutions Y or J are assigned to separate root classes.

That this definition makes sense is revealed in the following theorem.

Theorem 6.6: Equivalent point groups belong to the same root class.

Proof: By theorem 6.2, equivalent point groups are orthogonal equivalent. It follows that they have congruent root systems.

Therefore, root classes contain full equivalence classes of point groups.

A set M of involutions corresponds to every root system,

$$M := \{Z_1, \ldots, Z_s, -I, (Y_1, \ldots, Y_v, J_1, \ldots, J_w)\}.$$

Each root class is characterized through a point group of maximal order.

Definition 6.7: The point group of maximal order in a root class is called the holohedry of that root class.

The holohedry of a root class contains all involutions of M.

Theorem 6.7: Point groups belonging to a root class K are normal in the holohedry H of that root class.

Proof: Assume that the point group $P<H$ belongs to K. We have to show that for every symmetry operation $S \in H$ the relation $S^{-1}PS=P$ is fulfilled. Essential involutions $Y, J \in H$ are by definition 6.6 common to both P and H hence, they fulfil this relation. The same does the central inversion $-I \in H$. Now let $S:=Z_iZ_j$, $Z_i, Z_j \in M$. Because for every $Z_i \in M$ we have $Z_iPZ_i=P$ it follows that

$$Z_jZ_iPZ_iZ_j = Z_jPZ_j = P$$

which proves the above theorem.

Definition 6.8: Two involutions $Z_i, Z_j \in M$ are called related in P, $Z_i \sim Z_j$, if $Z_i Z_j = S \in P$.

This is an equivalence relation and we can classify the involutions of M into classes of related involutions. For spaces of even dimension the rotation $S_i := Z_i$ is an improper rotation. For spaces of odd dimension $S_i := -IZ_i$ is an improper rotation. Let P^+ be a point group of the first kind. By theorem 6.4, there exists a point group of the second kind

$$P = P^+ \cup P^+S_i.$$

If Z_i and Z_j are not related under P^+ then

$$P_1 := P^+ \cup P^+S_i \text{ and } P_2 = P^+ \cup P^+S_j$$

are non-equivalent point groups.

In what follows we derive the root classes and we will propose a possible nomenclature for the root classes.

In one-dimensional space there exists only one root class which we denote by the symbol "1". It contains the two one-dimensional point groups C_1 and C_i rspectively.

In two-dimensional space the number of root classes is infinite however, these may be grouped into a few series as is shown in Table 6.4.

Table 6.4 The root classes in two-dimensional space.

Root class	Crystal system	Point groups
1|1	Oblique	C_1, C_2
2|2	Rectangular	C_s, C_{2v}
3	Trigonal	C_3, C_{3v}
4	Square	C_4, C_{4v}
5	—	C_5, C_{5v}
6	Hexagonal	C_6, C_{6v}
.	—	.
.		.
m	—	C_m, C_{mv}

The symbol "1|1" indicates that there are two linearly independent directions, and that only the central inversion exists. The symbol "2|2" indicates that there are two involutions Z_1 and Z_2 which necessarily are perpendicular. We note that the symbol "2|1" does not represent a root class because the central inversion -I generates a second involution Z_2. The symbol "m", m∈ℕ, m>2, represents a principal plane (it is a two-dimensional space) with a rotation of order m. The rotation is generated by two involutions Z_1 and Z_2 having their root vectors in the principal plane. The root vectors enclose an angle of 180°/m.

It gives no difficulty to derive the root classes in three-dimensional space. The result is shown in Table 6.5.

Table 6.5 The root classes in three-dimensional space

Root class	Crystal system	Point groups
1|1|1	Triclinic	C_1, C_i
2|1|1	Monoclinic	C_2, C_s, C_{2h}
2|2|2	Orthorhombic	C_{2v}, D_2, D_{2h}
3|1	Trigonal	C_3, C_{3i}, C_{3v}, D_3, D_{3d}
4|2	Tetragonal	C_4, S_4, C_{4h}, C_{4v}, D_{2d}, D_4, D_{4h}
5|1	—	C_5, C_{5i}, C_{5v}, D_5, D_{5d}
6|2	Hexagonal	C_{3h}, C_6, C_{6h}, C_{6v}, D_{3h}, D_6, D_{6h}
.	—	.
.		.
m|1 (m odd)	—	C_m, C_{mi}, C_{mv}, D_m, D_{md}
m|2(m=4n+2)	—	C_m, $C_{\frac{1}{2}mh}$, C_{mh}, C_{mv}, $D_{\frac{1}{2}mh}$, D_m, D_{mh}
m|2 (m=4n)	—	C_m, S_m, C_{mh}, C_{mv}, $D_{\frac{1}{2}md}$, D_m, D_{mh}
O	Cubic	T, T_h, T_d, O, O_h
I	—	I, I_h
K	—	K, K_h

The bars between the numbers of a root class symbol indicate that the point groups belonging to this root class are reducible. The dimensions of the invariant subspaces can be deduced from these numbers. Thus the numbers 1 and 2 represent one-dimensional subspaces, the number m, m>2, represents a two-dimensional subspace. The symbols "O" and "I" denote root classes having irreducible point groups. In the symbols of reducible root classes a single odd number is not possible e.g. "2|2|1", "3|2", or "4|1" are not root classes. In up to three-dimensional space every crystal system corresponds to a full root class (for a definition of crystal system and crystal family the reader is referred to the International Tables, Vol. A). In higher dimensional spaces this is no longer true.

The root system belonging to the root class "O" is shown in Figure 6.7. The point group T has a congruent root system hence, it belongs to the same root class. The corresponding set of involutions fall, under T, into two classes of related involutions. One class of involutions corresponds to the twofold axes of the symmetry scaffolding of T. The other class corresponds to the root vectors which are determined through the intersections of the principal planes belonging to the threefold axes. Hence there exist two non-equivalent point groups of the second kind, viz.: T_h and T_d.

The root classes in four-dimensional space are given in Table 6.6 (the crystal families in four- dimensional space are taken from Brow1). There is no agreement between root classes and crystal families in four-dimensional space. The assignment of the four-dimensional point groups to the root classes is not yet accomplished. This will be the task of future research. Also a suitable nomenclature for point groups in higher dimensional spaces which shows the relation to the root classes is still missing. Proposals for a nomenclature of the crystal classes in four dimensional space were made by Whittacker (Whit1) and Weigel (Weig1).

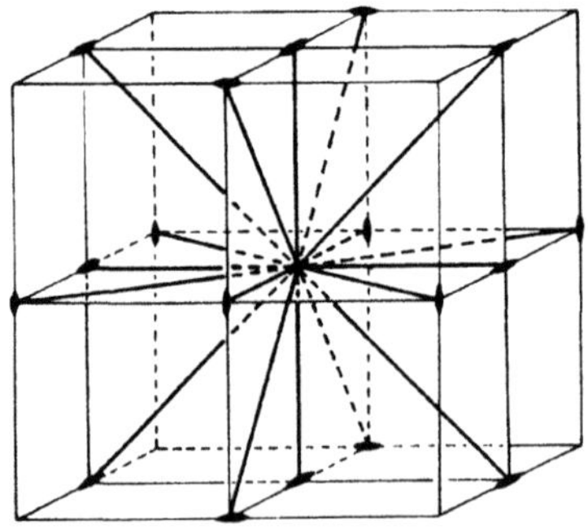

Figure 6.7. The root system belonging to the root class O

Table 6.6. The root classes of four-dimensional space

Root class		Crystal family
$1 \mid 1 \mid 1 \mid 1$		Hexaclinic
$2 \mid 1 \mid 1 \mid 1$		Triclinic
$\bar{1} \mid \bar{1} \mid 1 \mid 1$		Diclinic
$2 \mid 2 \mid 1 \mid 1$		Monoclinic
$2 \mid 2 \mid 2 \mid 2$		Orthogonal
$3 \mid 1 \mid 1$	<	Hexagonal monoclinic
$4 \mid 1 \mid 1$		Tetragonal monoclinic
$5 \mid 1 \mid 1$		——
$6 \mid 1 \mid 1$	<	Hexagonal monoclinic
.		.
$3 \mid 2 \mid 1$	<	Hexagonal orthogonal
$4 \mid 2 \mid 2$		Tetragonal orthogonal
$5 \mid 2 \mid 1$		——
$6 \mid 2 \mid 2$	<	Hexagonal orthogonal
.		.
$3 \mid 3$	<	Dihexagonal monoclinic diclinic orthogonal
$4 \mid 4$		Ditetragonal monoclinic diclinic orthogonal
$4 \mid 6$		Hexagonal tetragonal
$5 \mid 5$	><	Decagonal
$6 \mid 6$	<	Dihexagonal monoclinic diclinic orthogonal
$6 \mid 8$		——
.		
$7 \mid 7$		——
.		.
$8 \mid 8$	>	Octagonal
$8 \mid 10$		——
.		
.		
$10 \mid 10$	><	Decagonal
$10 \mid 12$		——
.		
.		
$12 \mid 12$	>	Dodecagonal
.		
.		
$h \mid k$		——

Table 6.6 (continued)

Root class		Crystal family
$\overline{3\vert 3}$	<	Diisohexagonal
$\overline{4\vert 4}$	<	Hypercubic
$\overline{5\vert 5}$	><	Icosahedral
$\overline{6\vert 6}$	<	Diisohexagonal
$\overline{7\vert 7}$		——
$\overline{8\vert 8}$	><	Hypercubic
$\overline{9\vert 9}$		——
$1\overline{0\vert 1}0$	><	Icosahedral
$1\overline{1\vert 1}1$		——
$1\overline{2\vert 1}2$	><	Diisohexagonal
.		
.		
$\overline{m\vert m}$		——
O|2		Cubic orthogonal
I|2		——
A_4^*	<	Icosahedral
F_4	<	Hypercubic
G_4		——
K_4		——

The symbol "<" indicates that the crystal family contains point groups which are not contained in the root class. The symbol ">" indicates that the root class contains point groups which are not contained in the crystal family. With K_4 we denote the root class belonging to the infinite point group of the sphere in four-dimensional space.

The root class "$\bar{1}|\bar{1}|1|1$" is determined by the central inversion $-I$, and the essential involution

$$J := \begin{pmatrix} -1 & 0 & 0 & 0 \\ 0 & -1 & 0 & 0 \\ 0 & 0 & 1 & 0 \\ 0 & 0 & 0 & 1 \end{pmatrix} .$$

The root class "$\overline{m|m}$" has an essential involution Y which interchanges the two principal planes. The holohedry in the root class "$\overline{m|m}$" has order $8m^2$ if m is even and $16m^2$ if m is odd.

6.6. Isomorphism types of point groups

The classification of crystal classes is based on the equivalence of their matrix representations in E^n. Alternatively we may classify the point groups according to their isomorphism type.

Definition 6.9: Two groups G and G' are isomorph and belong to the same isomorphism type if there is a bijective mapping of G onto G' which conserves the composition.

Example 6.3: In E^3 the point groups of order 2, viz.: C_i, C_s, and C_2, which correspond to a center of symmetry, a mirror plane, and a twofold axis respectively represent geometrically completely different symmetry scaffoldings. However, they are isomorph to the unique abstract group of order 2, viz.: $\{\alpha,\alpha^2=1\}$.

The results of this classification are summerized in the following theorem:

Theorem 6.7: There exist 2 in 1-dimensional, 9 in 2-dimensional, 18 in 3-dimensional and 117 isomorphism types of crystallographic point groups in 4-dimensional space.

The isomorphism types of the crystallographic point groups are listed in Table 6.8 for the 2-dimensional space and in Table 6.9 for the 3-dimensional space. In Table 6.7 are listed the numbers of non-isomorphic abstract groups of order h up to h=48.

6.7. Historical remarks

In 1815 the mineralogist Samuel Weiss (1780-1856) introduced the seven crystal systems (Weis1). This classification implicitly contained the crystallographic restriction which forbides a crystal symmetry of order five.

Table 6.7 Numbers of non-isomorphic groups of order h

h	N(h)	h	N(h)	h	N(h)	h	N(h)
1	1	13	1	25	2	37	1
2	1	14	2	26	2	38	2
3	1	15	1	27	5	39	2
4	2	16	14	28	4	40	14
5	1	17	1	29	1	41	1
6	2	18	5	30	4	42	6
7	1	19	1	31	1	43	1
8	5	20	5	32	51	44	4
9	2	21	2	33	1	45	2
10	2	22	2	34	2	46	2
11	1	23	1	35	1	47	1
12	5	24	15	36	14	48	52

Table 6.8 The isomorphism types of the 2-dimensional geometric crystal classes

isomorphism type	standard notation	geometric crystal classes
1.1	C_1	C_1
2.1	C_2	C_2, C_s
3.1	C_3	C_3
4.1	$C_2 \times C_2$	C_{2v}
4.2	C_4	C_4
6.1	$C_3 \times C_2$	C_6
6.2	D_6	C_{3v}
8.4	D_8	C_{4v}
12.3	$D_6 \times C_2$	C_{6v}

The 32 crystal classes in E^3 were determined for the first time in 1826 by the mineralogist Moritz Ludwig Frankenheim (1801-1872) (Fran1) as it was recently discovered by J.J. Burckhardt (Burc4). They were redetermined again in 1830 by the mineralogist Johann Friedrich Christian Hessel (1796-1872) (Hess1, Hess2). Frankenheim and Hessel both derived the 32 possible symmetries of polyhedra whose faces fulfil the law of rationality (cf. section 3.4). In the same year the mathematician Evariste Galois (1811-1832) published his treatise "Sur la théorie des nombres" in which he introduced the concept of a mathematical group which later became so important in crystallography (Galo1). The work of Frankenheim and Hessel remainded unnoticed and independently the crystal classes

Table 6.9 The isomorhism types of the 3-dimensional geometric crystal classes

isomorphism type	standard notation	geometric crystal classes
1.1	C_1	C_1
2.1	C_2	C_2, C_i, C_s
3.1	C_3	C_3
4.1	$C_2 \times C_2$	C_{2h}, D_2, C_{2v}
4.2	C_4	C_4, S_4
6.1	$C_3 \times C_2$	C_{3i}, C_6, C_{3h}
6.2	D_6	D_3, C_{3v}
8.1	$C_2 \times C_2 \times C_2$	D_{2h}
8.2	$C_4 \times C_2$	C_{4h}
8.4	D_8	D_4, C_{4v}, D_{2d}
12.1	$C_3 \times C_2 \times C_2$	C_{6h}
12.3	$D_6 \times C_2$	D_{3d}, D_6, C_{6v}, D_{3h}
12.5	A_4	T
16.6	$D_8 \times C_2$	D_{4h}
24.6	$D_6 \times C_2 \times C_2$	D_{6h}
24.10	$A_4 \times C_2$	T_h
24.15	S_4	O, T_d
48.36	$S_4 \times C_2$	O_h

were redetermined in 1849 by August Bravais (1811-1863) who however missed the rotoreflection (Brav1). A complete list was provided again in 1867 by Axel Gadolin (1828-1892) (Gado1). The crystallographic point groups were subsequently derived in various other ways. Non-crystallographic point groups were investigated by Werner Nowacki (Nowa1).

Four-dimensional groups were investigated by the mathematician Eduard Goursat (1858-1936) (Gour1). His paper contains three parts. In part II he determined four-dimensional point groups through direct products of two- and three-dimensional point groups. However, some of these groups have no real representation in four-dimensional space. In part III he generated four-dimensional point groups by reflections on the walls of a spherical tetrahedron. Point groups generated by reflections were further investigated by H.M.S. Coxeter whose book on regular poytopes has become standard (Coxe1). Four-dimensional groups were also derived using quaternions (for details the reader is referred to DuVa1). In 1949 Carl Hermann (1898-1961) described the symmetry operations in spaces of arbitrary dimensions (Herm3). The four-dimensional crystal classes were investigated by A.C. Hurley in 1951 (Hurl1, Hurl2). A complete list of four-dimensional crystal classes was compiled by Bülow in 1967. They are contained in the book of Harold Brown, Rolf Bülow, Joachim Neubüser, Hans Wondratschek, and Hans Zassenhaus (Brow1).

As already mentioned Weiss empirically derived the seven crystal systems. However, he used orthogonal axes for the triclinic and monoclinic system considering these to be subsystems of the orthogonal crystal system. This classification was based on the external shape of crystals. Gadolin emphasized that the point group symmetry determines the physical properties of a crystal. The tensor of a physical property remains invariant under a point group. It is by chance that in three-dimensional space the lattice symmetries are root class holohedries. In higher dimensional spaces this is no longer true. Unfortunately the holohedries in three-dimensional space were defined by Niggli as beeing lattice symmetries (Nigg3). For the definition of crystal systems in higher dimensions Neubüser, Wondratschek, and Bülow also used the lattice symmetries (Neub1). However, lattice symmetries have no physical significance. Because of the different definitions there is no longer a correspondence between root classes and crystal systems or crystal families in higher dimensional spaces.

7. Lattice Symmetries

We investigate in this chapter the action of a crystallographic point group P on a lattice T^n.

7.1. Definitions

In section 3.1 it was proved that a regular point system X consists of a finite set of congruent point lattices L_i each parallelly placed in space. Let $L \subset X$ be a point lattice in E^n. We choose a point $x_0 \in L$ as an origin O and then we identify each point $x_j \in L$ through its vector $\vec{x}_j$. Thus a lattice T^n is obtained. Among the lattice vectors of T^n we choose a lattice basis $\vec{a}_1, \ldots, \vec{a}_n$ (cf. section 3.2). We consider a crystallographic point group P which maps the lattice T^n onto itself. By theorem 5.1, the group P has a representation in E^n through unimodular nxn matrices with integral coefficients, if referred to a lattice basis. In section 6.1 the point groups were classified into geometric classes. A finer classification of the crystallographic point groups is obtained if we take into account the lattices T^n on which the groups act and the way a group acts on a lattice.

Definition 7.1: Two crystallographic point groups P and P' with lattices T^n and $T^{n\prime}$ are called arithmetically equivalent and belong to the same arithmetic crystal class if there exists an affine mapping φ which maps P onto P' and thereby maps T^n onto $T^{n\prime}$,

$$P' = \varphi P \varphi^{-1}, \quad T^{n\prime} = \varphi T^n.$$

Synonymous to arithmetic crystal class, $\mathbb{Z}$-class is used.

Because the point lattices $L_i \subset X$ are parallelly placed in space it follows that the symmetry of a single point lattice L is equal or higher than the symmetry of the regular point system X of which L is a subset only. Among all crystallographic point groups P_i which map the lattice T^n onto itself the one of highest order is distinguished to contain all origin preserving symmetry operations of the lattice T^n.

Definition 7.2: A Bravais point group B is a crystallographic point group which contains all isometries that map a lattice T^n onto itself.

In a similar way as we have classified the crystallographic point groups the Bravais point groups may be classified into geometric or arithmetic Bravais classes.

The Bravais point groups enable us to classify the lattices T^n.

Definition 7.3: Two lattices T^n and $T^{n'}$ with Bravais point groups B and B' belong to the same Bravais type of lattices if there exists an affine mapping φ which maps B onto B' and thereby maps T^n onto $T^{n'}$,

$$B' = \varphi B \varphi^{-1}, \quad T^{n'} = \varphi T^n.$$

By this definition there exists a one-to-one correspondence between the Bravais types of lattices and the arithmetic Bravais classes. In spaces of even dimension the central inversion $-I$ is a proper rotation hence, the lattice T^n may have a definite chirality. Because of physical reasons it is useful to consider only special affine transformations φ^+ which conserve the chirality of T^n. Thus more special Bravais types of lattices are distinguished (cf. section 8.1).

7.2. Bravais point groups

In what follows we characterize the Bravais point groups among the crystallographic point groups. In terms of the positive-definite n-nary quadratic form as described in section 4.1, a Bravais point group B corresponds to a group of linear substitutions which transform a quadratic form onto itself. Each such transformation corresponds to a symmetry operation of the corresponding lattice. We look at these symmetry operations in a geometric way.

Theorem 7.1: A Bravais point group B contains the central inversion $-I$.

Proof: From the definition of a lattice it follows that if $\vec{t}$ is a lattice vector then also $-\vec{t}$ is a lattice vector.

We consider a lattice T^n with Bravais point group B. Let $S\in B$ be a symmetry operation of T^n with symmetry support U (cf. section 5.4). Let d be the dimension of U with $0\leq d\leq n$. The space E^n is decomposed into U and its orthogonal complement V of dimension $h:=n-d$. E^n is the direct sum of U and V, $E^n=U+V$. Because the subspace U remains fixed under the symmetry operation S it follows that the subspace V is an invariant subspace under S.

<u>Theorem 7.2:</u> If the Bravais point group B of T^n contains a symmetry operation S of order k with d-dimensional symmetry support U, and if V , the orthogonal complement of U in E^n, then there exists in each of these subspaces a basis of lattice vectors $\vec{b}_i\in T^n$.

Proof: Let $\vec{a}_1,\ldots,\vec{a}_n$ be a lattice basis in T^n. For d=0 or d=n the above theorem is evident. Assume $0<d<n$. Among the basis vectors $\vec{a}_1,\ldots,\vec{a}_n$ we find at least d basis vectors, say $\vec{a}_1,\ldots,\vec{a}_d$ which have linearly independent non-zero components $\vec{u}_1,\ldots,\vec{u}_d$ in U and components $\vec{v}_1,\ldots,\vec{v}_d$ in V. Let $S\in B$ be of order k. We calculate

$$\vec{b}_i:=\vec{a}_i+S\vec{a}_i+\ldots+S^{k-1}\vec{a}_i$$

$$=k\vec{u}_i+\vec{v}_i+S\vec{v}_i+\ldots+S^{k-1}\vec{v}_i = k\vec{u}_i,$$

for $i=1,\ldots,d$. The $\vec{b}_i$ are lattice vectors in U which remain fixed under S because

$$S\vec{b}_i = S\vec{a}_i+S^2\vec{a}_i+\ldots+\vec{a}_i = \vec{b}_i.$$

Similarly, there exist at least $h:=n-d$ basis vectors $\vec{a}_j$, say $\vec{a}_{d+1},\ldots,\vec{a}_n$, which have linearly independent non-zero components $\vec{v}_j$ in V. We calculate

$$\vec{b}_j:= S\vec{a}_j-\vec{a}_j = S\vec{v}_j-\vec{v}_j$$

for $j=d+1,\ldots,n$. Because V is invariant under S it follows that $S\vec{v}_j\in V$. Hence the $\vec{b}_j$ are lattice vectors in V.

We note that in general the lattice vectors $\vec{b}_1,\ldots,\vec{b}_n$ do not form a lattice basis of T^n. From theorem 7.2 it follows that there exists in each of the subspaces U and V a sublattice $T^d\subset T^n$ and $T^h\subset T^n$ respectively which are orthogonal, that is, each lattice vector $\vec{q}\in T^d$ is orthogonal to every lattice vector $\vec{r} \in T^h$. However, it is not true, in general, that $T^n = T^d + T^h$.

As above let U be the symmetry support of $S \in B$ and V its orthogonal complement in E^n.

<u>Theorem 7.3:</u> If the Bravais point group B of T^n has a symmetry operation S of order $k>1$ and if the invariant subspace V has a decomposition into mutual orthogonal $\mathbb{Z}$-irreducible subspaces V_i then B contains an involution $Z := R_1 R_2 \ldots R_s$ as a generating symmetry operation.

Proof: Let $h := n-d$ be the dimension of V. By theorem 7.2, the subspace V has a basis of lattice vectors $\vec{b}_1, \ldots, \vec{b}_h$. Let $\vec{b}_1$ be a shortest lattice vector in V. We assume $\vec{b}_1$ has non-zero components in the subspaces $V_1, \ldots, V_v$, $v \le s$. Equivalent lattice vectors are obtained through the action of S^i on $\vec{b}_1$, $\vec{b}_i := S^{i-1}\vec{b}_1$, $i=2,\ldots,k$. Assume that among these h' are linearly independent and span the subspace V'. If $\dim(V') < \dim(V)$ then we take a shortest lattice vector say $\vec{b}_{h'+1}$ in the orthogonal complement of V' in V and determine the equivalent lattice vectors. We continue in this way until for $\vec{b}_h$ it holds that $V' = V$. Under the action of S^i the component $\vec{v}_{j1}$ in the principal plane π_j generates in π_j a regular m_j-gon as is shown in Figure 7.1. The m_j-gon has $C_{m_j v}$-symmetry. Therefore, there exists for each principal plane π_j a hyperplane H_j with plane normal $\vec{r}_j$ (root vector) in π_j such that the line of intersection $\sigma_j := H_j \cap \pi_j$ bisects $\vec{v}_{j1}$ and $\vec{v}_{j2}$. The reflection R_j in H_j maps $\vec{v}_{j1}$ onto $\vec{v}_{j2}$ and thereby maps π_j onto itself and leaves the orthogonal complement of $\vec{r}_j$ in E^n fixed (for a one-dimensional subspace V_i the root vector $\vec{r}_i$ is in V_i). By assumption the $\mathbb{Z}$-irreducible subspaces V_i are mutual orthogonal. It follows that the root vectors $\vec{r}_i$, $i=1,\ldots,s$, are mutual orthogonal hence, the product of the R_i is an involution,

$$Z := R_1 R_2 \ldots R_s .$$

The symmetry support W of Z is given through the intersection of all hyperplanes H_j, $j=1,\ldots,s$,

$$W = \bigcap_{j=1}^{s} H_j .$$

The action of Z on $\vec{b}_1$ is given by

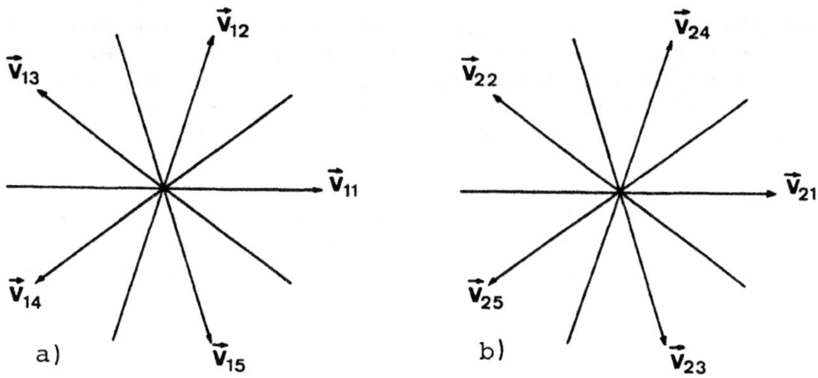

Figure 7.1. 10-gons generated in the principal planes π_1 and π_2 through the actions of a 5-fold rotation in 4-dimensional space.

$$Z\vec{b}_1 = Z\vec{v}_{11}+\ldots+Z\vec{v}_{s1} = \vec{v}_{12}+\ldots+\vec{v}_{s2} = \vec{b}_2.$$

Similarly, for all basis vectors $\vec{b}_i$ equivalent lattice vectors $\vec{b}_j=Z\vec{b}_i$ are found. Every lattice vector $\vec{q}\in T^h$ is written uniquely as $\vec{q}:=m_1\vec{b}_1+\ldots+m_h\vec{b}_h$, $m_i\in\mathbb{Z}$. It follows that

$$\vec{q}':= Z\vec{q} = m_1Z\vec{b}_1+\ldots+m_hZ\vec{b}_h = m_1\vec{b}_2+\ldots.+m_h\vec{b}_j$$

is a lattice vector in T^h. Every lattice vector $\vec{t}\in T^n$ is written uniquely as $\vec{t}=\vec{u}+\vec{v}$ with components $\vec{u}$ in U and $\vec{v}$ in V. We calculate

$$\vec{t}':= Z\vec{t} = Z\vec{u} + Z\vec{v} = \vec{u} + Z\vec{v}.$$

We have to determine $Z\vec{v}$. By theorem 7.2, there exists a basis of lattice vectors in the symmetry support W of Z and in its orthogonal complement in E^n. Therefore, we can determine $\vec{v}':=\vec{v}-\vec{d}$, $\vec{d}\in T^h$ such that

$$S\vec{v}'-\vec{v}' = S\vec{v}-\vec{v} - (S\vec{d}-\vec{d}) := \vec{w}\in T^h$$

is a lattice vector orthogonal to W. Therefore,

$$Z\vec{v}'-\vec{v}' = \vec{w} = Z\vec{v}-\vec{v} - (Z\vec{d}-\vec{d}).$$

Hence, $Z\vec{v}=\vec{v}+\vec{x}$, where $\vec{x}:=\vec{w}+Z\vec{d}-\vec{d}\in T^h$. It follows that

$$\vec{t}' = \vec{u} + \vec{v} + \vec{x} = \vec{t}+\vec{x}\in T^n.$$

Remark: In the case that two ore more $\mathbb{Z}$-irreducible subspaces have the same rotations then the principal planes within these subspaces are not uniquely defined. It follows that the root vectors need not be orthogonal and hence, there exists no involution Z.

Applying theorems 7.1 and 7.3 the Bravais point groups can be found among the crystallographic point groups and we classify them into geometric Bravais classes. In two-dimensional space the geometric Bravais classes are found to be

$$C_2,\ C_{2v},\ C_{4v},\ C_{6v}.$$

In three-dimensional space it is easily verified that

$$C_i,\ C_{2h},\ D_{2h},\ D_{3d},\ D_{4h},\ D_{6h},\ O_h$$

are geometric Bravais classes. In four-dimesnsional space the geometric Bravais classes were calculated by Bülow (Bülo1,Bülo2). The results of these calculations are summerized in the following theorem:

Theorem 7.4: There are 1 in 1-dimensional, 4 in 2-dimensional, 7 in 3-dimensional, and 33 geometric Bravais classes in 4-dimensional space.

7.3. Bravais types of lattices

For each geometric Bravais class a Bravais point group B is taken as a representative and we have to find the possible lattices which are invariant under B. The lattices are then classified according to their arithmetic Bravais point groups. Thereby, we have to consider for a given Bravais point group B affine transformations φ only which map B onto iteslf hence, φ belongs to the affine normalizer of B, $\varphi \in N_{/A}(B)$ (cf. section 8.3). Let $M:=\{S_1,\ldots,S_m\}$ be a set of generating symmetry operations of B (cf. section 6.1). For a symmetry operation $S_i \in M$ let $\tau(S_i)$ be the set of all lattices invariant under S_i.

<u>Theorem 7.5:</u> The set of all lattices $\tau(B)$ which are invariant under the Bravais point group B is given through

$$\tau(B) = \bigcap_{S_i \in M} \tau(S_i)$$

Proof: From $S_i\tau(S_i)=\tau(S_i)$ it follows that

$$\tau(S_i^m) = \tau(S_i).$$

From $S_i[\tau(S_i)\cap\tau(S_j)]=S_j[\tau(S_i)\cap\tau(S_j)]=S_iS_j[\tau(S_i)\cap\tau(S_j)]$ $=\tau(S_i)\cap\tau(S_j)$ it follows that

$$\tau(S_iS_j) = \tau(S_i)\cap\tau(S_j).$$

Since every symmetry operation $S\in B$ can be represented as a product

$$S = \prod_{S_i \in M} S_i$$

the above theorem follows.

For a symmetry operation $S_1\in B$ let $\{V_1,...,V_s\}$ be the set of $\mathbb{Z}$-irreducible subspaces of S_1. In what follows we determine the set of all lattices $\tau(S_1)$. To begin with we determine a lattice for a $\mathbb{Z}$-irreducible space V of dimension n. Let k be the order of the symmetry operation S_1 which acts on V.

<u>Theorem 7.6:</u> If the cyclic group C_k is $\mathbb{Z}$-irreducible then every non-trivial subgroup $C_h<C_k$ $(h>1)$ leaves only the origin fixed.

Proof: Because C_k is $\mathbb{Z}$-irreducible it leaves only the origin fixed. We can assume that k is not prime, $k=h\cdot l$. Suppose that $C_h<C_k$ $(h>1)$ leaves a subspace U of dimension $d>0$ fixed. The coset decomposition of C_k becomes

$$C_k = C_h \cup \rho_2C_h \cup...\cup \rho_mC_h.$$

As C_k is cyclic it follows that $\rho_iC_h\rho_i^{-1}=C_h$ hence, $\rho_iU=U$. Because ρ_i, $i=2,..,m$, leaves U fixed it follows that C_k leaves U fixed which is a contradiction.

By theorem 5.4, the dimension of V is $n=\psi(k)$. Since $(1-1/p_i)<1$ it follows that $n<k$. There exists a unique decomposition of V into principal planes π_j, $j=1,\ldots,s:=n/2$. Referred to a cartesian coordinate system $\vec{e}_1,\ldots,\vec{e}_n$ assumed within these pricipal planes, the symmetry operation S_1 has the following representation,

$$S_{1,0}:=\begin{pmatrix} \cos 2\pi h_1/k & -\sin 2\pi h_1/k & & & \\ \sin 2\pi h_1/k & \cos 2\pi h_1/k & & 0 & \\ & & \ddots & & \\ & 0 & & \cos 2\pi h_s/k & -\sin 2\pi h_s/k \\ & & & \sin 2\pi h_s/k & \cos 2\pi h_s/k \end{pmatrix},$$

where the integers $h_1<\ldots<h_s$ are less than $k/2$ and are relatively prime to k, that is $\gcd(h_i,k)=1$. By theorem 5.4, there exist exactly s such integers (cf. section 5.3). If $k=1$ or $k=2$ then $n=1$ and we can choose any vector $\vec{b}_1$ in V as a lattice vector. If $k>2$ then by theorem 7.3 the Bravais point group B also contains an involution Z as a generating symmetry operation. Let W be the symmetry support of Z. By theorem 7.2, W has a basis of lattice vectors. If $k=3,4$, or 6 then $n=2$ hence, there exists a single principal plane π. We can choose an arbitrary vector $\vec{b}_1\in W$ as a lattice vector. If $k=5$, or $k>6$ then V is decomposed into s principal planes π_j which, in this case, cannot contain a lattice vector because $2\cos 2\pi h_j/k$ is not integral. We have to choose a lattice vector $\vec{b}_1\in W$ such that $\vec{b}_1$ has non-zero components $\vec{v}_{j1}$ in each principal plane π_j because otherwise V^n would be $\mathbb{Z}$-reducible. Under the action of S^i lattice vectors $\vec{b}_i$ equivalent to $\vec{b}_1$ are obtained, $\vec{b}_i:=S^{i-1}\vec{b}_1$, $i=2,\ldots,k$. Because $n<k$ it follows that the $\vec{b}_i$ generate a lattice T^n. Depending on the the choice of the components $\vec{v}_{j1}$ the angle β between $\vec{b}_1$ and $\vec{b}_2$ is $360°/k<\beta<360°h_s/k$. Therefore, the $\vec{b}_i$ not necessarily are shortest lattice vectors of T^n. Only for $n\leq 2$ the $\vec{b}_i$ are always shortest lattice vectors. By theorem 3.12, a lattice basis $\vec{a}_1,\ldots,\vec{a}_n$ of T^n can be found through the construction of the Dirichlet parallelotope of T^n (for dimensions up to $n=30$ it was found that always $\vec{b}_1,\ldots,\vec{b}_n$ is a lattice basis of T^n). Using the transformation $A=(a_{ij})$ we express the basis vectors $\vec{a}_1,\ldots,\vec{a}_n$ referred to the cartesian coordinate system $\vec{e}_1,\ldots,\vec{e}_n$,

$$\begin{pmatrix} \vec{a}_1 \\ \cdot \\ \cdot \\ \cdot \\ \vec{a}_n \end{pmatrix} = \begin{pmatrix} a_{11} \dots a_{1n} \\ \cdot \quad \cdot \\ \cdot \quad \cdot \\ \cdot \quad \cdot \\ a_{n1} \dots a_{nn} \end{pmatrix} \begin{pmatrix} \vec{e}_1 \\ \cdot \\ \cdot \\ \cdot \\ \vec{e}_n \end{pmatrix} .$$

In reference to the lattice basis $\vec{a}_1, \dots, \vec{a}_n$, the symmetry operation S_1 has a representation with integer coefficients (cf. section 3.1),

$$S_{1,1} := A \circ S_{1,0} A^t .$$

If $\vec{b}_1, \dots, \vec{b}_n$ are taken as a lattice basis then the representation of S_1 has the form

$$S_{1,1} = \begin{pmatrix} 0 & 0 & 0 & \dots 0 & \lambda_1 \\ 1 & 0 & 0 & & \cdot \\ 0 & 1 & 0 & & \cdot \\ 0 & 0 & 1 & & \cdot \\ \cdot & & \cdot & & \cdot \\ \cdot & & & \cdot & \cdot \\ 0 & \dots & & 0 \; 1 & \lambda_n \end{pmatrix} .$$

Up to dimension n=30 it was found that λ_i only assumes the integers −1,0,1. It is not proved that this is true in general. If k is prime then $\lambda_i = -1$, $i = 1, \dots, n$. All lattices obtained in this way are equivalent under the affine normalizer $N_{/A}(B)$.

We now assume that S_1 has two $\mathbb{Z}$-irreducible subspaces V_1 and V_2 of dimensions h_1 and h_2. The symmetry operation S_1 is represented, referred to a cartesian coordinate system, by

$$S_{1,0} := \begin{pmatrix} R_1 & 0 \\ & \\ 0 & R_2 \end{pmatrix} .$$

Let k_1, k_2 be the orders of R_1 and R_2 respectively with $k_1 < k_2$. It follows that the order of S_1 is $k = (k_1 k_2)/\gcd(k_1, k_2)$. Two cases have to be considered:

1. We can determine a lattice basis in each subspace V_i as described above.

2. Another lattice basis which, in general, is not equivalent to the first one under $N_{/A}(B)$, is obtained if the

lattice vector $\vec{b}_1$ has non-zero components in both subspaces V_1 and V_2. Then necessarily $\vec{b}_1$ must have non-zero components in each principal plane because otherwise one or both of the subspaces V_i would be $\mathbb{Z}$-reducible. Equivalent lattice vectors are obtained through the action of S^i, $\vec{b}_i := S^{i-1}\vec{b}_1$, $i=2,...,k$. Among these h' are linearly independent and span a subspace V'. If dim(V')<dim(V) then we take a shortest lattice vector in the orthogonal complement of V' in E^n and determine again the equivalent lattice vectors.

We note that if $k_1 \neq k_2$ then S^k leaves the subspace V_1 fixed hence, V_1 and V_2 are orthogonal and by theorem 7.2 there exists a basis of lattice vectors in V_1 and V_2. The corresponding sublattices T^{h_1} and T^{h_2} are mutual orthogonal but this does not imply that $T^n = T^{h_1} + T^{h_2}$. If $k_1 = k_2$ then the principal planes are pairwaise not uniquely determined. In this case there exist invariant sublattices T^{h_1} and T^{h_2} which are not orthogonal to each other.

If there are more than two $\mathbb{Z}$-irreducible subspaces additional combinations occur. For instance, the lattice vector $\vec{b}_1$ may have non-zero components in one, two, or more $\mathbb{Z}$-irreducible subspaces.

The next step is to determine for each lattice as determined above the set of all lattices which are invariant under $S \in M$. In order to do this we consider the n^2-dimensional space Γ of all nxn matrices. In the space Γ the positive-definte n-nary quadratic forms are contained within some cone. Since the quadratic forms are symmetric it follows that Γ can be reduced to Γ' of dimension $n(n+1)/2$. However, the following results are more conveniently formulated if we use Γ instead of Γ'. Each quadratic form $C=(c_{ij})$ corresponds to a lattice basis of a lattice T^n. The quadratic form C is identified through a point $\gamma \in \Gamma$. The coordinates ξ_k of γ, $k=1,...,n^2$, are defined by $\xi_k := c_{ij}$, where $k=(i-1)n+j$. Equivalent quadratic forms C' of T^n correspond to equivalent points $\gamma' \in \Gamma$. Thus a lattice T^n corresponds to an infinite set of equivalent points in Γ. Within this set of equivalent points by theorem 4.1 one point γ_0 is distinguished to correspond to the reduced form. Thus, the lattice T^n is uniquely represented by γ_0.

Because the symmetry operation S is an isometry it has to leave a quadratic form C invariant (cf. section 1.3),

$$C = S^t C S.$$

We rewrite this equation in the form,

$$\gamma = S^t x S^t \gamma,$$

where $S^t x S^t$ is the direct product (Kronecker product) of S^t with itself. It has a representation through a $n^2 x n^2$ matrix. The form C is considered as a point $\gamma \in \Gamma$. Thus $\sigma := S^t x S^t$ is a mapping of Γ onto itself and γ is a fixed point of σ. Let U be the symmetry support of σ. It corresponds to the set of all nxn matrices which are invariant under a certain representation of S. Under a change of lattice basis with transformation A, the quadratic form C is transformed according to $C' := ACA^t$. An equivalent lattice basis gives rise to an equivalent symmetry support $U' := AxAU$. Thus we obtain an infinite set of symmetry supports U_i which we call a star. Among the symmetry supports U_i of a star a symmetry support U_0 is chosen which contains at least one reduced quadratic form. A lattice basis which is not equivalent under $N_{/A}(B)$ gives rise to another star in Γ. Therefore, the set of all lattices $\tau(S)$ which are invariant under S is represented in Γ through the union of a finite number of stars. Among each star we have to take the symmetry support U_0.

In section 5.4 an arithmetic method was used to calculate the symmetry support of a symmetry operation S. We now give an alternative method which is based on representation theory.

<u>Theorem 7.7:</u> For a symmetry operation S of order k the dimension d of the symmetry support of $\sigma := S^t x S^t$ is given by

$$d = \frac{1}{k} \sum_{i=1}^{k} tr^2(S^i). \qquad (7.1)$$

Proof: Through the direct product $\sigma := S^t x S^t$ we get a representation of S in Γ. The dimension d of the symmetry support U of σ is equal to the number of times the identical irreducible representation occurs in the representation of σ. We use the orthogonality relation between the irreducible representations in order to obtain the dimension d_ν corresponding to the ν-th irreducible representation,

$$d_v = \frac{1}{h} \sum_{\varphi_i \in P} \chi_v(\varphi_i)\ tr(\varphi_i),$$

where h is the order of P. (For details the reader is referred to Brud1). Since the characters of the identical representation are $\chi_1(\varphi_i)=1$, and because $tr(S^t x S^t)=tr^2(S)$ the formula 7.1 follows. By theorem 5.2, the trace is invariant under a change of basis hence, d is an invariant property of S.

We determine $\tau(S_i)$ for each $S_i \in M$. By theorem 7.5, the set of all lattices $\tau(B)$ which are invariant under B is obtained through the intersection of the $\tau(S_i)$. We then have to eliminate those solutions which are equivalent under $N_{/A}(B)$. There may exist lattices $T^n \in \tau(B)$ whose Bravais point group B' is of higher order than B, B'>B, then T^n is non-characteristic for the generating Bravais point group B. We only take lattices which are characteristic in their Bravais point group.

In order to obtain the relations between the coefficients c_{ij} of the quadratic form C we calculate the projection operator P_1 corresponding to the identical irreducible representation (cf. section 5.3),

$$P_1 = \frac{1}{h} \sum_{S_i \in B} \sigma_i,$$

where h is the order of B. For a certain representation of B in Γ let U be its symmetry support. The projection operator P_1 projects the basis vectors of Γ into the invariant subspace U. Thus the required relations are obtained.

Finally we determine for the set of all lattices $\tau(B)$ the number N of free parameters which can be chosen arbitrarely in order to determine the coefficients c_{ij} of a quadratic form. Let $\sigma := S^t x S^t$, $S \in B$, be a representation of B in Γ and let U be its symmetry support. The dimension d of U is given by theorem 7.7,

$$d = \frac{1}{h} \sum_{S_i \in B} tr^2(S_i),$$

where h is the order of B. Since the quadratic form C is symmetric it follows that $N \leq d$. The sum $c_{ij}+c_{ji}$ is invariant under the transposition of C. Hence, if c_{ij} and c_{ji} contain different free parameters then only the sum of each such pair is a free parameter in Γ'. The number N can be calculated if we include the transposition of C to the set of the σ_i. Let B' be the group thus generated. We obtain N through

$$N = \frac{1}{h} \sum_{\sigma_i \in B'} tr(\sigma_i).$$

Example 7.1: In two-dimensional space E^2 the Bravais types of lattices for each geometric Bravais class are easily obtained. The four geometric Bravais classes are considered in turn.

1. The geometric Bravais class C_2. The generating symmetry operation of C_2 is the central inversion $-I$. The direct product $\sigma := (-I) \times (-I)$ is the identity matrix in Γ. Therefore, the symmetry support of σ is Γ. We can choose any point $\gamma \in \Gamma$ which belongs to a positive-definite quadratic form. The number of free parameters is three. Thus the quadratic form C has the form

$$C := \begin{vmatrix} a & c \\ c & b \end{vmatrix} .$$

It belongs to the oblique lattice. In order that C is positive-definite we have to require that $a,b>0$ and $ab-c^2>0$.

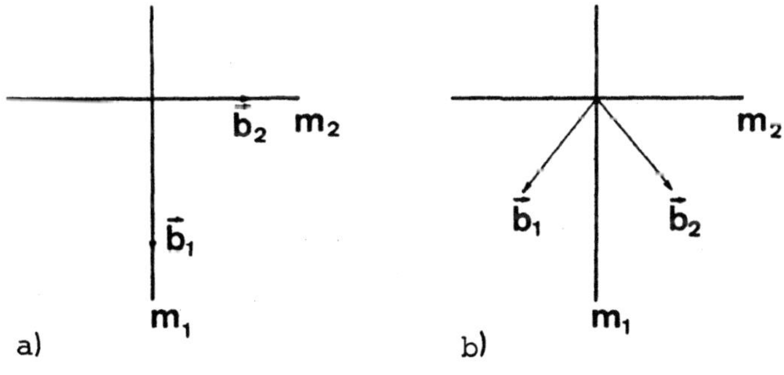

Figure 7.2. Alternative choices of basis vectors in the Bravais class C_{2v}

2. The geometric Bravais class C_{2v}. Let $M := \{S_1, S_2\}$ be the set of generating symmetry operations. A possible choice for S_1 and S_2 are the two reflections in the mirror lines m_1 and m_2 as shown in Figure 7.2. We choose a cartesian coordinate system by assuming $\vec{e}_1$ to

be in m_1 and $\vec{e}_2$ to be in m_2. Referred to this coordinate system we have the following representation of S_1 and S_2:

$$S_{1,0} := \begin{vmatrix} 1 & 0 \\ 0 & -1 \end{vmatrix}, \qquad S_{2,0} := \begin{vmatrix} -1 & 0 \\ 0 & 1 \end{vmatrix}.$$

First we consider S_1. It has two invariant one-dimensional subspaces denoted by $V_1:=m_1$ and $V_2:=m_2$ which, by theorem 7.2, are orthogonal. We obtain a lattice basis by choosing the vectors $\vec{b}_1$ in V_1 and $\vec{b}_2$ in V_2. The transformation is

$$\begin{vmatrix} \vec{b}_1 \\ \vec{b}_2 \end{vmatrix} = \begin{vmatrix} 1 & 0 \\ 0 & 1 \end{vmatrix} \begin{vmatrix} \vec{e}_1 \\ \vec{e}_2 \end{vmatrix}.$$

It follows that the representation of S_1 referred to this lattice basis becomes

$$S_{1,1} = \begin{vmatrix} 1 & 0 \\ 0 & -1 \end{vmatrix}.$$

The direct product $\sigma:=S^t x S^t$ becomes

$$\sigma_{1,1} = \begin{vmatrix} 1 & 0 & 0 & 0 \\ 0 & -1 & 0 & 0 \\ 0 & 0 & -1 & 0 \\ 0 & 0 & 0 & 1 \end{vmatrix}.$$

The symmetry support of $\sigma_{1,1}$ is two-dimensional,

$$U_{11} = \lambda_1 \ (1\ 0\ 0\ 0) + \lambda_2 \ (0\ 0\ 0\ 1).$$

Another non-equivalent lattice basis is obtained if we choose the vector $\vec{b}_1$ such that it has non-zero components $\vec{v}_1$ in V_1 and $\vec{v}_2$ in V_2. The basis vectors $\vec{b}_1$ and $\vec{b}_2:=S_1\vec{b}_1$ form a lattice basis. The transformation becomes

$$\begin{vmatrix} \vec{b}_1 \\ \vec{b}_2 \end{vmatrix} = \begin{vmatrix} v_1 & v_2 \\ v_1 & -v_2 \end{vmatrix} \begin{vmatrix} \vec{e}_1 \\ \vec{e}_2 \end{vmatrix}.$$

Referred to this lattice basis the representation of S_1 becomes

$$S_{1,2} = \begin{vmatrix} 0 & 1 \\ 1 & 0 \end{vmatrix}.$$

Its direct product becomes

$$\sigma_{1,2} = \begin{vmatrix} 0 & 0 & 0 & 1 \\ 0 & 0 & 1 & 0 \\ 0 & 1 & 0 & 0 \\ 1 & 0 & 0 & 0 \end{vmatrix} .$$

The corresponding symmetry support U_{12} is two-dimensional,

$$U_{12} = \lambda_1 \ (1 \ 0 \ 0 \ 1) + \lambda_2 \ (0 \ 1 \ 1 \ 0).$$

Similarly, for S_2 we get the following non-equivalent representations,

$$S_{2,1} := \begin{vmatrix} -1 & 0 \\ 0 & 1 \end{vmatrix} , \qquad S_{2,2} := \begin{vmatrix} 0 & 1 \\ 1 & 0 \end{vmatrix} .$$

As above we get the symmetry supports of $\sigma_{2,1}$ and $\sigma_{2,2}$, viz.: $U_{21}=U_{11}$ and $U_{22}=U_{12}$. Both contain a reduced form. We now determine the intersections of these symmetry supports. The following solutions are obtained:

1. $U_{11} \cap U_{21} = U_{11}$
2. $U_{12} \cap U_{22} = U_{12}$
3. $U_{11} \cap U_{22} = \lambda_1(1 \ 0 \ 0 \ 1)$.

The number of free parameters N is

$$N=d=1/4(4+0+4+0) = 2.$$

For the first solution we obtain a projection operator P_1,

$$P_1 := \tfrac{1}{2} \begin{vmatrix} 1 & 0 & 0 & 0 \\ 0 & 0 & 0 & 0 \\ 0 & 0 & 0 & 0 \\ 0 & 0 & 0 & 1 \end{vmatrix} .$$

It follows that $a:=c'_{11}=\tfrac{1}{2}c_{11}$, $b:=c'_{22}=\tfrac{1}{2}c_{22}$, and $c'_{12}=0$. The quadratic form becomes

$$C_1 := \begin{vmatrix} a & 0 \\ 0 & b \end{vmatrix} .$$

It belongs to the rectangular lattice. For the second solution the projection operator P_2 becomes

$$P_2 := \frac{1}{2}\begin{pmatrix} 1 & 0 & 0 & 1 \\ 0 & 1 & 0 & 0 \\ 0 & 0 & 1 & 0 \\ 1 & 0 & 0 & 1 \end{pmatrix}.$$

It follows that $a := c'_{11} = c'_{22} = \frac{1}{2}(c_{11} + c_{22})$ and $b := c'_{12} = \frac{1}{2}c_{12}$. The quadratic form becomes

$$C_2 := \begin{vmatrix} a & b \\ b & a \end{vmatrix}.$$

It belongs to the c-centered rectangular lattice. The third solution is one-dimensional only. It is invariant under both types of representations. The quadratic form becomes

$$C_3 := \begin{vmatrix} a & 0 \\ 0 & a \end{vmatrix}.$$

It is a limiting case of C_1 and C_2 and corresponds to a square lattice.

3. The geometric Bravais class C_{4v}. As generating symmetry operations we choose the fourfold rotation and the reflection in the mirror line m. Referred to a cartesian coordinate system we have the following representations:

$$S_{1,0} := \begin{vmatrix} 0 & -1 \\ 1 & 0 \end{vmatrix}, \qquad S_{2,0} := \begin{vmatrix} -1 & 0 \\ 0 & 1 \end{vmatrix}.$$

The symmetry operation S_1 is $\mathbb{Z}$-irreducible. We can choose an arbitrary basis vector $\vec{b}_1$ in m. An equivalent basis vector is obtained under the action of S_1, $\vec{b}_2 := S_1\vec{b}_1$. They form a lattice basis. The transformation is

$$\begin{vmatrix} \vec{b}_1 \\ \vec{b}_2 \end{vmatrix} = \begin{vmatrix} 1 & 0 \\ 0 & 1 \end{vmatrix} \begin{vmatrix} \vec{e}_1 \\ \vec{e}_2 \end{vmatrix}.$$

It follows that $S_{1,1} = S_{1,0}$. We calculate the direct product σ_{11} and determine its symmetry support,

$$U_{11} = \lambda_1\ (1\ 0\ 0\ 1).$$

The symmetry supports corresponding to S_2 were determined under 2., viz.:

$$U_{21} = \lambda_1(1\ 0\ 0\ 0) + \lambda_2(0\ 0\ 0\ 1)$$

$$U_{22} = \lambda_1(1\ 0\ 0\ 1) + \lambda_2(0\ 1\ 1\ 0).$$

We determine the intersections of these symmetry supports. The following solutions are obtained:

1. $U_{11} \cap U_{21} = U_{11}$
2. $U_{11} \cap U_{22} = U_{11}$.

The number of free parameters N is

$$N=d=1/8(4+0+4+0+0+0+0+0) = 1.$$

For both solutions we obtain the same projection operator P_1,

$$P_1 := \tfrac{1}{2}\begin{pmatrix} 1 & 0 & 0 & 1 \\ 0 & 0 & 0 & 0 \\ 0 & 0 & 0 & 0 \\ 1 & 0 & 0 & 1 \end{pmatrix}.$$

It follows that $a:=c'_{11}=c'_{22}=\frac{1}{2}(c_{11}+c_{22})$. The quadartic form becomes

$$C_1 := \begin{vmatrix} a & 0 \\ 0 & a \end{vmatrix}.$$

It belongs to the square lattice. Both solutions are equivalent under $N_{/A}(B)$. One lattice is rotated by 45° against the other.

4. The geometric Bravais class C_{6v}. As generating symmetry operations we choose the sixfold rotation and the reflection in the mirror line m. Referred to a cartesian coordinate system we have the following representation,

$$S_{1,0} := \begin{vmatrix} 1/2 & -\sqrt{3}/2 \\ \sqrt{3}/2 & 1/2 \end{vmatrix}, \qquad S_{2,0} := \begin{vmatrix} -1 & 0 \\ 0 & 1 \end{vmatrix}.$$

The symmetry operation S_1 is $\mathbb{Z}$-irreducible. We can choose an arbitrary basis vector $\vec{b}_1$ in m. The basis vector $\vec{b}_2$ is obtained by $\vec{b}_2 := S_1\vec{b}_1$. The transformation is

$$\begin{vmatrix} \vec{b}_1 \\ \vec{b}_2 \end{vmatrix} = \begin{vmatrix} 1 & 0 \\ 1/2 & \sqrt{3}/2 \end{vmatrix} \begin{vmatrix} \vec{e}_1 \\ \vec{e}_2 \end{vmatrix}.$$

Referred to this lattice basis the representation of S_1 becomes

$$S_{1,1} := \begin{vmatrix} 0 & -1 \\ 1 & 1 \end{vmatrix} .$$

We calculate the direct product $\sigma_{1,1}$,

$$\sigma_{1,1} := \begin{pmatrix} 0 & 0 & 0 & 1 \\ 0 & 0 & -1 & 1 \\ 0 & -1 & 0 & 1 \\ 1 & -1 & -1 & 1 \end{pmatrix} .$$

Its symmetry support is

$$U_{11} = \lambda_1(2 \ 1 \ 1 \ 2).$$

The symmetry supports corresponding to S_2 were determined above under 2., viz.:

$$U_{21} = \lambda_1(1 \ 0 \ 0 \ 0) + \lambda_2(0 \ 0 \ 0 \ 1)$$
$$U_{22} = \lambda_1(1 \ 0 \ 0 \ 1) + \lambda_2(0 \ 1 \ 1 \ 0).$$

The intersection of these symmetry supports becomes

$$U_{11} \cap U_{22} = U_{11}.$$

The number of free parameters is

$$N=d=1/12(4+1+1+4+0+0+0+0+0+0) = 1.$$

From the projection operator P_1,

$$P_1 := \tfrac{1}{6} \begin{vmatrix} 4 & -2 & -2 & 4 \\ 2 & -1 & -1 & 2 \\ 2 & -1 & -1 & 2 \\ 4 & -2 & -2 & 4 \end{vmatrix} ,$$

we get the relations $a := c'_{11} = c'_{22} = \frac{1}{6}(4c_{11} - 2c_{12} - 2c_{21} + 4c_{22})$, $c'_{12} = \frac{1}{6}(2c_{11} - c_{12} - c_{21} + 2c_{22}) = \frac{1}{2}c'_{11}$ hence, the quadratic form becomes

$$C := \begin{vmatrix} a & \frac{1}{2}a \\ \frac{1}{2}a & a \end{vmatrix} .$$

It belongs to the hexagonal lattice.

Altogether there are five Bravais types of lattices in two-dimensional space. The procedure to obtain them is straight forward however, in higher dimensions large matrices have to be handled therefore, the use of a

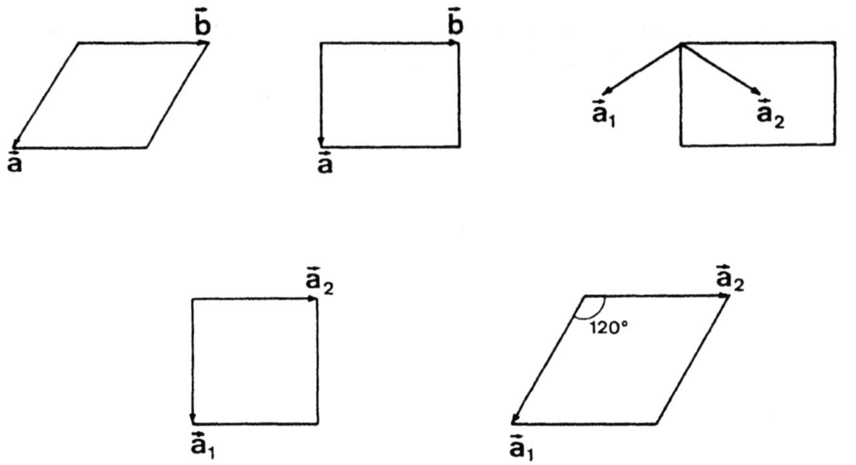

Figure 7.3. The five Bravais types of lattices in two-dimensional space E^2

computer is inevitable. The Bravais types of lattices in three-dimensional space were determined geometrically by Frankenheim (Frank2) and Bravais (Brav2). In four-dimensional space the Bravais types of lattices were determined by calculating the subgroups of the maximal groups of 4x4 integral matrices (Bülo2). In five-, and six-dimensional space first the geometric Bravais classes were determined and then splitted into $\mathbb{Z}$-classes (Ples2, Ples3). The results of these calculations are summerized in the following theorem:

Theorem 7.8: There are 1 in 1-dimensioanl, 5 in 2-dimensional, 14 in 3-dimensional, 64 in 4-dimensional, 189 in 5-dimensional, and 826 Bravais types of lattices in 6-dimensional space.

7.4. Arithmetic crystal classes

The arithmetic crystal classes arising from a geometric crystal class K can be determined in the same way as the Bravais types of lattices. Indeed there is a one-to-one correspondence between the Bravais types of lattices and the arithmetic Bravais classes. For a given crystallographic point group $P \in K$ as a representative, the admissible lattices are determined up to equivalence under the affine normalizer $N_{/A}(P)$.

However, knowing the arithmetic Bravais classes in E^n, group-subgroup relations can be used to find all arithmetic crystal classes. A representative of an arithmetic Bravais class is given through an ordered pair (B,T^n) of a Bravais point group B and a lattice T^n on which B acts. Let $B_s \geq B$ be the Bravais point group of T^n. If $B_s > B$, then T^n is called non-characteristic with respect to B. Otherwise if $B_s = B$, then T^n is called characteristic. In what follows we consider for B both characteristic and non-characteristic lattices (cf. section 7.3).

Every crystallographic point group P can be assigned to a minimal Bravais point group B, $P \leq B$, such that there is no other Bravais point group B' between P and B. For a certain lattice $T^n \in \tau(B)$ with Bravais point group B_s we split up P into arithmetic crystal classes. The non-equivalence of two arithmetic crystal classes geometrically corresponds to a different setting of the symmetry scaffolding of P with respect to the lattice T^n. The group of all symmetry operations which map the point group P onto itself is the normalizer of P in the orthogonal group O(n), $N_O(P)$, (cf. section 8.3). Therefore, we have to compare the normalizers $N_O(P)$ and $N_O(B_s)$. Two cases have to be considered:

1. If $N_O(P) \geq N_O(B_s)$, then T^n determines a single arithmetic crystal class of P.

2. If $N_O(P) < N_O(B_s)$, then necessarily the index $m := [N_O(B_s) : N_O(P)]$ is finite and we determine the coset decomposition,

$$N_O(B_s) = \rho_1 N_O(P) \cup \ldots \cup \rho_m N_O(P),$$

   where ρ_1 is the identity. In this case each lattice $\rho_i T^n$, $i=1,\ldots,m$, determines a non-equivalent arithmetic crystal class of P.

We do this step for a representative T^n of each Bravais type of lattices belonging to B. Therewith all arithmetic crystal classes arising from P are obtained.

Example 7.2: In the plane E^2 let T^2 be a hexagonal lattice with Bravais point group C_{6v}. The orthogonal normalizer of C_{6v} is C_{12v}. We consider the following subgroups belonging to C_{6v}, viz.: C_3, C_{3v}, C_6. Their orthogonal normalizers are $C_{\infty v}$, C_{6v}, and $C_{\infty v}$ respectively. Because $C_{\infty v}>C_{12v}$ it follows that to C_3 and C_6 belongs a single arithmetic crystal class. On the other hand $C_{6v}<C_{12v}$ of index 2. It follows that two non-equivalent arithmetic crystal classes arise out of C_{3v}. For the second class the lattice is rotated by 30° with respect to the symmetry scaffolding of C_{3v} as shown in Figure 7.4.

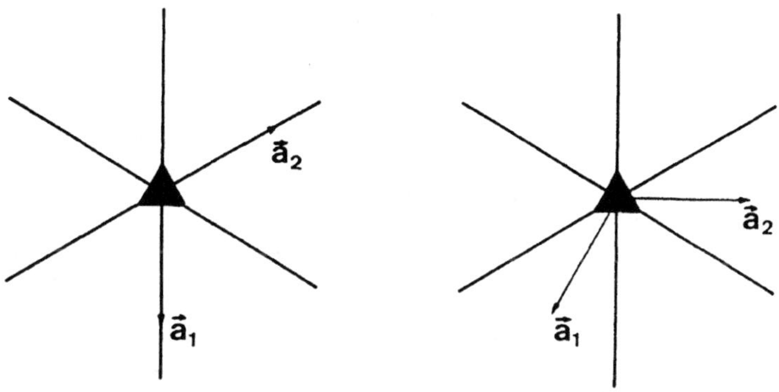

Figure 7.4. The two settings of a hexagonal lattice with respect to the symmetry scaffolding of C_{3v}

The arithmetic crystal classes in three-dimensional space were determined by Niggli and Nowacki (Nigg4). These correspond to the tree-dimensional symmorphic space-group types determined by Fedorov (Fedo2, Fedo3). The arithmetic crystal classes in four-dimensional space were calculated by Bülow (Bülo1,Bülo2). The results of these calculations are summerized in the following theorem:

<u>Theorem 7.9:</u> There are 2 in 1-dimensional, 13 in 2-dimensional, 73 in 3-dimensional and 710+70 arithmetic crystal classes in 4-dimensional space.

In spaces of dimension $n\geq 4$ pairs of enantiomorphic arithmetic crystal classes occur (cf. section 1.3). For $n=4$ the notation 710+70 indicates that there are 70 enantiomorphic pairs.

7.5. Crystal forms

According to the correspondence rule the facets of a crystal are parallel to lattice planes (cf. section 3.4). The growth of a crystal facet with facet vector $\vec{f}$, in general, is not the same in the positive or in the negative direction of $\vec{f}$. Therefore, we assign by $\vec{f}$ an orientation to a set of parallel lattice planes. By theorem 3.3, the facet vector $\Delta\vec{f}$ has integral components, if referred to the reciprocal lattice basis. A facet vector $\vec{f}_1$, together with a finite positive real number ϑ, determine a hyperplane H_1^0 having distance ϑ from the origin. H_1^0 separates the space into two half-spaces H_1^+ and H_1^-, where H_1^+ is assumed to contain the origin (cf. section 2.1). Under the action of a crystallographic point group P_0 facet vectors $\vec{f}_i$ equivalent to $\vec{f}_1$ are generated, $\vec{f}_i:=S_i\vec{f}_1$, $S_i\in P_0$. Each facet vector $\vec{f}_i$, together with the distance ϑ, determine an equivalent half-space H_i^+.

Definition 7.4: A crystal form is the intersection of all half-spaces H_i^+ which are equivalent under a crystallographic point group P_0.

A crystal form is called open or closed if it is an unbounded or a bounded set respectively. The stabilizer P_s of a crystal form (cf. section 1.4) may be of higher order than the generating point group P_0, $P_s \geq P_0$. A crystal form is called characteristic or non-characteristic with respect to P_0 if $P_s=P_0$ or $P_s>P_0$ respectively.

With the following definition crystal forms are classified into types of crystal forms:

Definition 7.5: Two crystal forms F and F', with stabilizers P_s and P_s' respectively, are called equivalent and belong to the same type of crystal forms if F is combinatorially equivalent to F' and if P_s is equivalent to P_s'.

Let $St_0(\vec{f}_1)<P_0$ be the stabilizer of the facet vector $\vec{f}_1$. The crystal form generated by $\vec{f}_1$ is called general or special in P_0 if $St_0(\vec{f}_1)=C_1$ or $St_0(\vec{f}_1)>C_1$ respectively. Let the d-dimensional subspace U, $0\leq d\leq n$, be the symmetry support of $St_0(\vec{f}_1)$. Let S^{n-1} be a sphere of radius $|\vec{f}_1|$. We say that the facet vector $\vec{f}_1$ has (d-1) degrees of freedom in $S':=U\cap S^{n-1}$. Suppose now that d>1. We vary the facet vector $\vec{f}_1$ by small amounts $\delta\vec{f}\in S'$ and then verify if the type of crystal forms has changed. In this way S' is divided into fields of existence each having the same type of crystal forms. Two neighbouring (d-1)-dimensional fields of existence are separated by a (d-2)-dimensional field of existence which corresponds to a limiting type of crystal forms.

In three-dimensional space we observe that a limiting type of crystal forms always exhibits a higher symmetry than the types of crystal forms in both adjacent fields of existence. It is not proved that this is true in higher dimensions.

The following method, based on this observation, allows us to find in a systematic way all types of crystal forms occuring in a crystallographic point group P_0 (cf. section 8.5). The way to do it is to determine for P_0 all non-characteristic crystal forms. Let P_1 be a supergroup of P_0 of index k and let $St_0(\vec{f}_1)$ and $St_1(\vec{f}_1)$ be the stabilizers of $\vec{f}_1$ in P_0 and P_1 respectively. The facet vector $\vec{f}_1$ belongs to a non-characteristic crystal form only if the index $[St_1(\vec{f}_1):St_0(\vec{f}_1)]$ is equal to the index k because otherwise P_1 would generate more facet vectors than P_0. Each pair of subgroups $St_0(\vec{f})<P_0$ and $St_1(\vec{f})<P_1$ which satisfies this condition is a solution for a non-characteristic crystal form. If there are no subgroups which fulfil this condition then P_1 and any supergroup $P_2>P_1$ do not occur as a stabilizer of a crystal form generated by P_0. For each facet vector $\vec{f}$ there exists a maximal supergroup $P_s \geq P_0$ which still fulfils the condition. Thereby, we restrict ourselves to crystallographic point groups because, for non-crystallographic point groups, the facet vectors do not have rational components. Moreover, we need consider only supergroups which are non-equivalent under the orthogonal normalizer $N_O(P_0)$.

The results of these calculations are summerized in the following theorem:

<u>Theorem 7.10:</u> There are 2 in 1-dimensional, 9 in two-dimensional, and 47 types of crystal forms in three-dimensional space.

Example 7.3: We consider in three-dimensional space the crystallographic point group T (cf. section 6.3). We choose the conventional lattice basis having basis vectors in the three mutual orthogonal twofold axis.

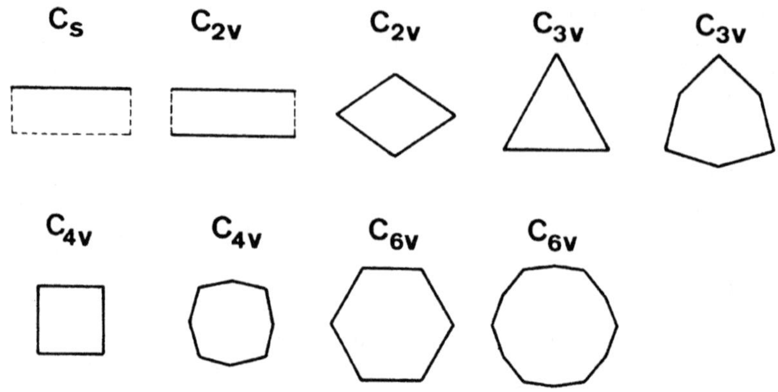

Figure 7.5. The 9 types of crystal forms in two-dimensional space

Characteristic crystal forms:

The generating facet vector is $\vec{f}_1:=(h,k,l)$, $h,k,l\in\mathbb{Z}$. The crystal form thus generated is the tetrahedral pentagondodecahedron (cf. Figure 7.6 and Table 7.1).

Non-characteristic crystal forms:

1. Supergroup T_h of index 2. Solutions are given by the horizontal mirror plane. Facet vector $\vec{f}_1:=(h,k,0)$, $St_0(\vec{f}_1)=C_1$, $St_1(\vec{f}_1)=C_s$, $[C_s:C_1]=2$. The crystal form thus generated is the pentagondodecahedron.

2. Supergroup T_d of index 2. Solutions are given by the diagonal mirror plane. Facet vector $\vec{f}_1:=(h,h,l)$, $St_0(\vec{f}_1)=C_1$, $St_1(\vec{f}_1)=C_s$, $[C_s:C_1]=2$. the crystal form thus generated is the deltoiddodecahedron. Another solution is given by the threefold axis. Facet vector $\vec{f}_1:=(1,1,1)$, $St_0(\vec{f}_1)=C_3$, $St_1(\vec{f}_1)=C_{3v}$, $[C_{3v}:C_3]=2$. The form thus generated is the triakistetrahedron.

3. Supergroup O of index 2. All solutions are contained in O_h.

4. Supergroup O_h of index 4. A solution is given by the fourfold axis. Facet vector $\vec{f}_1:=(1,0,0)$, $St_0(\vec{f}_1)=C_2$, $St_1(\vec{f}_1)=C_{4v}$, $[C_{4v}:C_2]=4$. The form thus generated is the cube. Another solution is given by the diagonal twofold axis. Facet vector $\vec{f}_1:=(1,1,0)$, $St_0(\vec{f}_1)=C_1$, $St_1(\vec{f}_1)=C_{2v}$, $[C_{2v}:C_1]=4$. The form thus generated is the rhomb-dodecahedron.

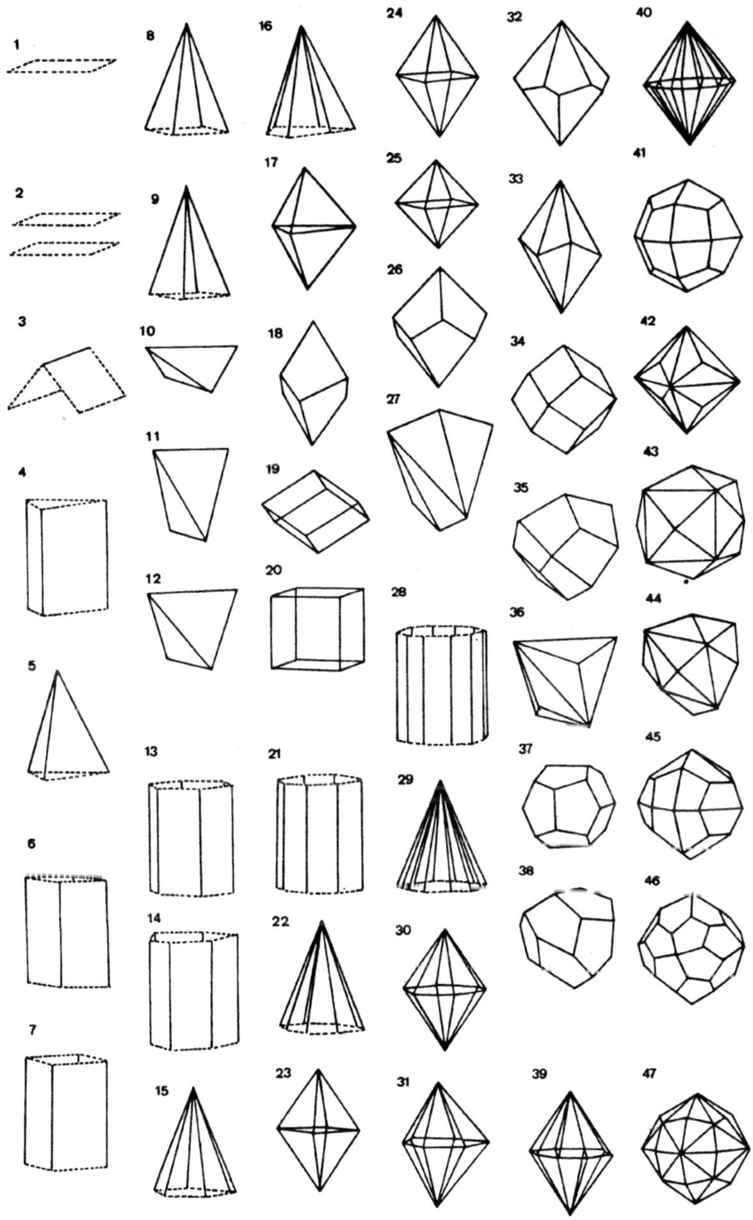

Figure 7.6. The 47 types of crystal forms in three-dimensional space

Table 7.1. The 47 types of crystal forms in three-dimensional space

No.	number of facets	name after R.L. Parker	stabilizer
1	1	pedion	$C_{\infty v}$
2	2	pinacoid	$D_{\infty h}$
3	2	sphenoid	C_{2v}
4	3	trigonal prism	D_{3h}'
5	3	trigonal pyramid	C_{3v}
6	4	orthorhombic prism	D_{2h}
7	4	tetragonal prism	D_{4h}
8	4	tetragonal pyramid	C_{4v}
9	4	orthorhombic pyramid	C_{2v}
10	4	orthorhombic bisphenoid	D_2
11	4	tetragonal bisphenoid	D_{2d}
12	4	tetrahedron	T_d
13	6	hexagonal prism	D_{6h}
14	6	ditrigonal prism	D_{3h}
15	6	hexagonal pyramid	C_{6v}
16	6	ditrigonal pyramid	C_{3v}
17	6	trigonal bipyramid	D_{3h}
18	6	trigonal trapezohedron	D_3
19	6	rhombohedron	D_{3d}
20	6	hexahedron (cube)	O_h
21	8	ditetragonal prism	D_{4h}
22	8	ditetragonal pyramid	C_{4v}
23	8	orthorhombic bipyramid	D_{2h}
24	8	tetragonal bipyramid	D_{4h}
25	8	octahedron	O_h
26	8	tetragonal trapezohedron	D_4
27	8	tetragonal scalenohedron	D_{2d}
28	12	dihexagonal prism	D_{6h}
29	12	dihexagonal pyramid	C_{6v}
30	12	hexagonal bipyramid	D_{6h}
31	12	ditrigonal bipyramid	D_{3h}
32	12	hexagonal trapezohedron	D_6
33	12	ditrigonal scalenohedron	D_{3d}
34	12	rhomb-dodecahedron	O_h
35	12	deltoiddodecahedron	T_h
36	12	triakistetrahedron	T_d
37	12	pentagondodecahedron	T_h
38	12	tetrahedral pentagon-dodecahedron	T
39	16	ditetragonal bipyramid	D_{4h}
40	24	dihexagonal bipyramid	D_{6h}
41	24	deltoidikositerahedron	O_h
42	24	triakistetrahedron	O_h

Table 7.1. (continued)

No.	number of facets	name	stabilizer
43	24	tetrakishexahedron	O_h
44	24	hexakistetrahedron	T_d
45	24	diakisdodekahedron	T_h
46	24	pentagonikositetrahedron	O
47	48	hexakisoctahedron	O_h

5. The non-crystallographic supergroup I_h of index 10. A solution is given by the fivefold axis. Facet vector $\vec{f}_1:=((1+\sqrt{5})/2,1,0)$, $St_0(\vec{f}_1)=C_1$, $St_1(\vec{f}_1)=C_{5v}$, $[C_{5v}:C_1]=10$. The form thus generated is the regular dodecahedron. It is no crystal form because the facet vector has irrational components.

If we remove the law of rationality then additional forms occur as was shown in Example 7.3.5. At all 56 crystallographic and non-crystallographic forms occur within the 32 crystal classes which were determined by Galiulin (Gali1).

7.6. Historical remarks

The characterization of space lattices through ternary quadratic forms was pointed out in 1831 by Ludwig Seeber (Seeb2). In 1842 Moritz Ludwig Frankenheim (1801-1872) determined fifteen types of space lattices (Frank2). One type of lattice he listed twice. A complete derivation was later given by Bravais (Brav2). These types of lattices are now called the Bravais types of lattices. The Bravais types of lattices in four-dimensional space were geometrically investigated in 1958 by Alen Lindsay Mackay and G. Stuart Pawley (Mack1). Additions and corrections to their list by A.M. Zamorzaev and B.V. Tsekinovskii and by N.V. Belov (1891-1982) and T.S. Kuntsevich still did not gave the final result. A complete list was compiled by Rolf Bülow in 1967 (Bülo1). The Bravais types of lattices in five-dimensional space were compiled in 1980 by Wilhelm Plesken (Ples2) and in 1984 the Bravais types of lattices in six-dimensioal space were compiled by Plesken and Hanrath (Ples3).

The symmorphic space groups, which are in one-to-one correspondence to the arithmetic crystal classes, were derived in three-dimensional space by Fedorov. The term arithmetic crystal classes and its significance for the derivation of space groups was indroduced in 1934 by Johann Jakob Burckhardt (Burc1). Paul Niggli and Werner Nowacki determined the arithmetic crystal classes in three-dimensional space.

Already René Just Haüy (1743-1822) observed that the crystal forms could be related by integral proportions to a "primitive crystal form". In 1816 Christian Samuel Weiss (1780-1856) showed how crystal faces could be referred to a system of axes. He discovered that the intercepts of different crystal faces relative to the crystal axes are related by small whole numbers (Weis2). The indices used now are inversely proportional to Weiss' indices. They were proposed in 1825 by the mineralogist W. Whewell. A systematic description of crystal forms was performed in 1830 by Carl Friedrich Naumann (1797-1873). The present nomenclature of the crystal forms is a slight modification, due to R.L. Parker, of the one given by Paul Groth (Grot1). A different nomenclature was proposed by Martin Buerger (Buer1).

8. Space Groups

Up till now symmetry groups were considered only which leave at least one point fixed. This restriction is now removed and we investigate groups of symmetry operations which act transitively on a regular point system $X \subset E^n$.

8.1. Definitions

Space groups were defined in section 1.4 as groups of symmetry operations acting transitively on a regular point system $X \subset E^n$. The space groups are classified into space group types according to the following definition:

Definition 8.1: Two space groups G and G' belong to the same space group type if there exists an affine mapping φ which maps G onto G',

$$G' = \varphi G \varphi^{-1}.$$

Because of physical reasons it is useful to consider only affine transformations φ^+ of the special affine group $\mathbb{A}^+(n)$, which conserve the chirality of the coordinate system. For dimensions $n>2$ the number of special space group types thus obtained is larger than the number of space group types.

Definition 8.2: An orbit $O(H,x_0)$ is the set of all points equivalent to a point $x_0 \in E^n$ under a group H of symmetry operations,

$$O(H,x_0) := \{x_i \mid x_i = \sigma_i x_0,\ \sigma_i \in H\}.$$

If H is a space group G then by definition 1.3 an orbit $O(G,x_0)$ corresponds to a regular point system $X \subset E^n$. Let $St(x_0) < H$ be the stabilizer of the point x_0 in H (cf. definition 1.4). An orbit $O(H,x_0)$ is called general or special if $St(x_0) = C_1$ or $St(x_0) > C_1$ respectively, where C_1 is the trivial group containing the identity only.

Definition 8.3: A fundamental region F of a space group G is any simply connected region which contains from each orbit of G exactly one point.

Synonymous to fundamental region also asymmetric unit is used.

Since each orbit $O(G,x_0)$ corresponds to a regular point system there exists for $O(G,x_0)$ a maximal interstitial ball of radius R. It follows that the Dirichlet domain $D(x_0)$ is bounded. Remembering the properties 2.1 to 2.3 of Dirichlet domain partitions (cf. section 2.3) we see that the Dirichlet domain of a general orbit can be used to define a fundamental region of G. The following relations hold:

$$\mathrm{int}D(x_0) \subset F, \quad F \subset \mathrm{cl}D(x_0).$$

Thus only part of the boundary of $\mathrm{cl}D(x_0)$ belongs to the fundamental region F of G. Since $D(x_0)$ is bounded it follows that F is bounded. Fundamental regions for the two- and three-dimensional space groups are given in the International Tables, Vol. A (Hahn1).

In definition 6.2 the symmetry scaffolding of a point group P was defined. It is useful to define in a similar way the symmetry scaffolding of a space group G. However, as G is an infinite group, some special considerations are necessary. By definition 5.3 the symmetry support of a general symmetry operation $(S,\vec{s})\in G$ was defined as a coset $\vec{p}^{\perp}+U$. The number of symmetry operations which have an identical symmetry support may be infinite in the case that $d>0$ hence, we choose one symmetry operation among them for which S has maximal order k and for which the screw vector

$$\vec{t} = S^{k-1}\vec{s}+\ldots+S\vec{s}+\vec{s}$$

has minimal length. Such a symmetry operation we call a minimax symmetry operation. Let $\vec{t}'$ be a shortest lattice vector such that $\vec{t}=m\vec{t}'$. For a minimax symmetry operation it follows that $0\leq m<k/2$. If $d=1$ and $m>0$ it may be necessary to distinguish between right- and left-handed screw rotations. Let $U^{\perp}$ be the orthogonal complement of U. If there exists a vector $\vec{v}_1\in U^{\perp}$ such that the vectors $\vec{v}_i:=S^{i-1}\vec{v}_1$, $i=1,\ldots,n-1$, form a basis of $U^{\perp}$ then $(S,\vec{s})$ has a definite chirality. Let the nxn matrix A be defined by the components of $\vec{v}_1,\ldots,\vec{v}_{n-1},\vec{t}$. The minimax symmetry operation $(S,\vec{s})$ is called a right- or a left-handed screw rotation if $\det(A)>0$ or $\det(A)<0$ respectively (cf. sections 1.3 and 5.5). We can assign to each symmetry support a double weight $(\pm m,k)$, where + or − represents a right- or a left-handed screw rotation respectively. This allows us to define the symmetry scaffolding of a space group.

<u>Definition 8.4:</u> The symmetry scaffolding of a space group G is the double weighted union of the symmetry supports of all minimax symmetry operations $(S,\vec{s})\in G\setminus T^n$.

8.2. Derivation of space groups

By theorem 3.2, every space group G in n-dimensional Euclidean space E^n contains n linearly independent translations which generate a free abelian translation group T^n. Let P be a crystallographic point group which maps T^n onto itself. The ordered pair (P,T^n) determines an arithmetic crystal class (cf. section 7.1). We ask for the possible space groups which can be derived from (P,T^n).

8.2.1. Derivation by the congruences of Frobenius

The coset decomposition of G with respect to T^n is

$$G = T^n \cup T^n(S_2,\vec{s}_2) \cup T^n(S_3,\vec{s}_3) \cup \ldots ,$$

where the shift vectors $\vec{s}_i$ can be chosen modulo T^n. For the identity operation always $(I,0)$ is assumed. In order to construct a space group G it is necessary and sufficient to specify for each coset representative $(S_i,\vec{s}_i)$ a shift vector $\vec{s}_i$ such that $(S_i,\vec{s}_i)\in G$. This choice is not arbitrary. Using a geometric argument a direct method to determine the shift vectors is given in the third part of this section.

Two properties of space groups are revealed by the following theorems.

<u>Theorem 8.1:</u> The translation group T^n is normal in G and is a maximal free abelian subgroup of G.

Proof: T^n is a free abelian group. It is normal in G because for any symmetry operation $(S,\vec{s})\in G$ and any translation $\vec{t}\in T^n$ we obtain

$$(S,\vec{s})(I,\vec{t})(S,\vec{s})^{-1} = (I,S\vec{t}).$$

By requirement $S\in P$ maps T^n onto itself hence, $S\vec{t}\in T^n$ for every $\vec{t}\in T^n$. Suppose $T^n \subset G'$, where $G'<G$ is a larger

abelian subgroup of G. For any $(S,\vec{s})\in G'$ not contained in T^n we obtain as above

$$(S,\vec{s})(I,\vec{t})(S,\vec{s})^{-1} = (I,S\vec{t}).$$

Because G' is abelian we also have

$$(S,\vec{s})(I,\vec{t})(S,\vec{s})^{-1} = (I,\vec{t})$$

hence, $(I,S\vec{t})=(I,\vec{t})$ for all $\vec{t}\in T^n$. Since T^n has n linearly independent translations it follows that S=I hence, $(S,\vec{s})\in T^n$. Thus T^n is a maximal abelian subgroup of G.

Theorem 8.2: The quotient group G/T^n is isomorphic to a crystallographic point group P.

Proof: Suppose that $(S_i,\vec{s}_i)$ and $(S_j,\vec{s}_j)$ are in the same coset of G with respect to T^n. It follows that there exists a translation $\vec{t}\in T^n$ such that

$$(S_i,\vec{s}_i) = (S_j,\vec{s}_j)(I,\vec{t}).$$

From

$$(S_j,\vec{s}_j)^{-1}(S_i,\vec{s}_i) = (I,\vec{t})$$

it follows that $S_j^{-1}S_i=I$ hence, $S_j=S_i$. Inversely $(S_i,\vec{s}_i)$ and $(S_i,\vec{s}_j)$ are in the same coset because

$$(S_i,\vec{s}_i)(S_i,\vec{s}_j)^{-1} = (I,-\vec{s}_j+\vec{s}_i)\in T^n.$$

Therefore, there is a one-to-one correspondence between the cosets and the symmetry operations $S_i\in P$.

By theorem 3.1, a crystallographic point group P is finite hence, the index of T^n in G is finite.

Let h be the order of P. The quotient group $G/T^n\simeq P$ requires that the group is closed under coset multiplication,

$$T^n(S_i,\vec{s}_i)T^n(S_j,\vec{s}_j) = T^n(S_k,\vec{s}_k),$$

where $S_k=S_iS_j$, $i,j=1,\ldots,h$. It follows that

$$T^n(S_iS_j,S_i\vec{s}_j+\vec{s}_i) = T^n(S_k,\vec{s}_k)$$

hence, the translation parts on both sides of the above equation have to be equal up to a translation $\vec{t}_{ij}\in T^n$. Thus we get a Frobenius congruence (Frob1),

$$S_i\vec{s}_j+\vec{s}_i-\vec{s}_k = 0 \text{ modulo } T^n.$$

A similar congruence is set up for each pair $S_i,S_j\in P$. The shift vector of the identity operation always is assumed to be zero. A set of shift vectors $\vec{s}_1,\ldots,\vec{s}_h$ which fulfils all h^2 congruences is called a solution of the Frobenius congruences and gives rise to a possible space group. If $\vec{r}_1,\ldots,\vec{r}_h$ is another solution then also the sum of both solutions $\vec{s}_1+\vec{r}_1,\ldots,\vec{s}_h+\vec{r}_h$ is a solution. Particularly the set of shift vectors $\vec{s}_1=\ldots=\vec{s}_h=0$ is always a solution, called the zero-solution.

Definition 8.5: The space group corresponding to the zero-solution of the Frobenius congruences is called a symmorphic space group.

There is a one-to-one correspondence between the arithmetic crystal classes and the symmorphic space group types.

8.2.2. Derivation by group extension

Space groups can be considered as extensions of a crystallographic point group P by a lattice T^n. In extension theory the property of theorem 8.2 is expressed as a short exact sequence of group homomorphisms,

$$0 \longrightarrow T^n \xrightarrow{\alpha} G \xrightarrow{\beta} P \xrightarrow{\gamma} 1,$$

where 0 and 1 are the trivial groups for the additively written group T^n and the multiplicatively written group P respectively. The sequence is called exact because the image of each homomorphism is isomorphic to the kernel of its successor. Thus we have

$$\text{image}(\alpha) \simeq T^n = \text{kernel}(\beta)$$
$$\text{image}(\beta) \simeq P = \text{kernel}(\gamma),$$

where α is injective and β is surjective. It follows that $G/T^n\simeq P$.

In extension theory a Frobenius congruence is written as

$$S_i\vec{s}_j+\vec{s}_i-\vec{s}_k = \vec{t}_{ij} = f(S_i,S_j).$$

The function $f(S_i,S_j)$ is called a factor and has values in T^n. It is independent of the choice of origin. The normalizing condition is

$$f(I,S_i) = f(S_i,I) = 0.$$

A set of factors $f(S_i,S_j)$, $i,j=1,...,h$ is called a normalized 2-cocycle. It is a solution of the Frobenius congruences and corresponds to a mapping f: $P\times P \rightarrow T^n$. The set of all 2-cocycles is denoted by $Z^2(P,T^n)$. They form an additive abelian group. The 2-cocycles depend on the choice of coset representatives. If we change the representatives,

$$\vec{s}_i' := \vec{s}_i+\vec{t}_i,\ \vec{s}_j' := \vec{s}_j+\vec{t}_j,\ \vec{s}_k' := \vec{s}_k+\vec{t}_k \quad (\vec{t}_i,\vec{t}_j,\vec{t}_k \in T^n),$$

then a 2-cocyle changes by a 2-coboundary,

$$S_i\vec{t}_j+\vec{t}_i-\vec{t}_k.$$

The set of all 2-coboundaries is denoted by $B^2(P,T^n)$. It is a subgroup of $Z^2(P,T^n)$. Each element of the second cohomology group,

$$H^2(P,T^n) := Z^2(P,T^n)/B^2(P,T^n),$$

gives an extension G of P by T^n. The extension theory provides no procedure to determine the shift vectors $\vec{s}_i$ such that the Frobenius congruences are fulfilled for each pair $S_i,S_j \in P$. Only the rather weak result is known that the origin can be chosen such that the shift vectors are multiples of 1/h, where h is the order of the crystallographic point group P (Frob1). We note that in higher dimensions the order h can become very large.

8.2.3. Derivation by orbit splitting

The above theory cannot explain the remarkable observation that the number of space groups derived from an arithmetic crystal class may vary drastically from one class to another. The following geometric construction gives a basic insight into the structure of space groups and allows a direct determination of the shift vectors $\vec{s}_i$.

Let Σ be a symmorphic space group with crystallographic point group P and lattice T^n. We consider a general orbit

$O(\Sigma,x_0)$ which we represent as the union of all translates of an orbit of P (cf. Figure 1.1),

$$O(\Sigma,x_0) = \bigcup_{\vec{t}_i \in T^n} O(P,x_0)+\vec{t}_i .$$

Each space group G, derived from (P,T^n), we consider as a splitting of the orbit $O(P,x_0)$ in the following way:

Definition 8.6: A partition of an orbit $O(P,x_0)$ into k disjoint subsets M_i, $i=1,\dots,k$, is called a regular splitting of $O(P,x_0)$ if there exists a point group $P'<P$ which acts transitively on the points of a subset M_i and another point group $C'<P$ which acts transitively on the k subsets M_i. The group C' is called the complement of P' in P.

Theorem 8.3: A necessary and sufficient condition for a regular splitting of $O(P,x_0)$ is that at least one of the groups P' or C' is normal in P and that P' and C' only have the identity operation in common.

Proof: Let h be the order of P. Suppose P' is normal in P of index k. The coset decomposition of P with respect to P' becomes,

$$P = P'W_1 \cup \dots \cup P'W_k ,$$

where W_1 is the identity operation. Because P' is normal in P the cosets $P'W_i$ form a group under coset multiplication which is isomorphic to C'. The group C' is not uniquely determined in P. It depends on the choice of coset representatives. Under P' the orbit $O(P,x_0)$ splits up into k different orbits $O(P',x_i)$, with $x_i=W_ix_0$. Each orbit $O(P',x_i)$ corresponds to a subset M_i as introduced above and C' acts transitively on the k subsets M_i. By requirement P' and C' have only the identity operation in common hence, the subsets are disjoint. For each C' there is the possibility that $O(P,x_0)$ splits up under C' into h/k different orbits $M_i:=O(C',x_i)$, with $x_i=R_ix_0$, $R_i \in P'$. In this case P' acts transitively on the h/k subsets M_i. Therefore, the action of P' and C' on $O(P,x_0)$ can be interchanged.

The importance of the regular splitting of $O(P,x_0)$ is revealed in the following theorem:

Theorem 8.4: Every space group G derived from an arithmetic crystal class (P,T^n) corresponds to a regular splitting of $O(P,x_0)$.

Proof: Suppose G has a maximal point group $P'<P$ of index k. Under P' the orbit $O(P,x_0)$ splits up into k different orbits $O(P',x_i)$, with $x_i:=W_i x_0$, where W_i is a coset representative of P with respect to P'. All translates $O(P',x_i)+\vec{t}$, $\vec{t}\in T^n$, generate an orbit $O(\Sigma',x_i)$ of a symmorphic space group Σ' corresponding to (P',T^n). In order to obtain a space group G we have to place translates $O(\Sigma',x_i)+\vec{s}_i$, $i=1,...,k$, in space in such a way that their union forms a regular point system X in E^n,

$$X:=\bigcup_{i=1}^{k} O(\Sigma',x_i)+\vec{s}_i,$$

on which G acts transitively. This implies that all sets $O(\Sigma',x_i)+\vec{s}_i$ have to be disjoint and congruent to a regular point system X'. Hence, the cosets $P'W_i$, $i=1,...,k$, have to form a group under coset multiplication which is isomorphic to $C'<P$. It follows that at least P' or C' have to be normal in P hence, the splitting of $O(P,x_0)$ is regular.

Following Weissenberg, the point group P' is called a principal point group of G (cf. section 8.5).

Theorem 8.5: If $P',C'<P$ determine a regular splitting of $O(P,x_0)$ then the shift vectors of a space group G derived from an arithmetic crystal class (P,T^n) are among the translations of T'/T^n, where T' is the translation subgroup of N which is defined as

$$N:=N_{/A}(\Sigma') \text{ if } P' \text{ is normal in } P, \text{ or}$$
$$N:=N_{/A}(F') \text{ if } C' \text{ is normal in } P, \text{ or}$$
$$N:=N_{/A}(\Sigma') \cap N_{/A}(F') \text{ if } P' \text{ and } C' \text{ are normal in } P.$$

Proof: Suppose $P'<P$ is of index k. The coset decomposition of G with respect to T^n is

$$G = T^n \cup T^n(S_2,\vec{s}_2) \cup \ldots \cup T^n(S_h,\vec{s}_h),$$

where h is the order of P. We are free to choose the coset representatives, belonging to P', to have shift vectors $\vec{s}=0$ thus defining the origin to lie on the symmetry support of P'. The union of these cosets form a symmorphic space group $\Sigma'<G$. The coset representatives belonging to C', appart of the identity operation, all have shift vectors $\vec{s}\neq 0$ modulo T^n. Because C' is a group the union of these cosets form a space group $F'<G$. Since C' is the complement of P' in P the space group G is generated by the union of Σ' and F'. By requirement G acts transitively on

$$X := \bigcup_{i=1}^{k} O(\Sigma', x_i)+\vec{s}_i .$$

Suppose P' is normal in P. The symmetry operations of F' have to map the symmetry scaffolding of Σ' onto itself. This implies that F' is a subgroup of the affine normalizer (cf. section 8.3),

$$F'< N_{/A}(\Sigma').$$

Since $F' \cup \Sigma'$ generates G it follows that

$$G< N := N_{/A}(\Sigma').$$

The symmetry scaffolding of Σ' is mapped onto itself by the translations of N. By assumption P' is normal in P therefore, the symmetry scaffolding of Σ' is invariant under P hence, it is invariant under $C'<P$. Thus, the translations of N determine the possible shift vectors of F'. Suppose now that C' is normal in P. The same arguments show that

$$G< N := N_{/A}(F').$$

The translations of N map the symmetry scaffolding of F' onto itself. Suppose that both P' and C' are normal in P then

$$G< N_{/A}(\Sigma') \text{ and } G< N_{/A}(F')$$

hence,

$$G< N := N_{/A}(\Sigma') \cap N_{/A}(F').$$

The translations of N map the symmetry scaffolding of Σ' and of F' respectively onto itself. Let T' be the translation subgroup of N. Two translations $\vec{s}_1$, $\vec{s}_2 \in T'$ such that

$$\vec{s}_1+ \vec{s}_2= \vec{t} \in T^n,$$

give rise to equivalent space groups. Therefore, we can restrict ourselves to the translations of the quotient group T'/T^n.

Among the translations of T'/T^n only those have to be considered which are not equivalent under the normalizer of P in the general linear group $GL(n,\mathbb{Z})$.

For decomposable groups the group N can have infinitly small translations. Suppose P is decomposable. There exists a symmetry operation $S \in P$ such that the symmetry support U of S is an invariant subspace of P. Suppose the group N has infinitely small translations in U. By theorem 7.2, there exists a basis of lattice vectors in U. Let m be the order of S. By theorem 5.8, there exists a representation of $(S,\vec{s})$ such that $\vec{s}$ has rational components with denominator m. Particularly if P',C' define a regular splitting of $O(P,x_0)$ and $S \in C'$ then $\vec{s}$ can be chosen such that

$$\vec{s} = \vec{t}/k,$$

where k is the index of P' in P and $\vec{t}$ is a lattice vector in U.

If P',C' determine a regular splitting of $O(P,x_0)$ then some of the shift vectors $\vec{s} \in T'/T^n$ may determine a space group G which belongs to another regular splitting of $O(P,x_0)$. This can be seen as follows: Suppose that $W \in C'$, $W \neq I$, has shift vector $\vec{v}$ modulo T^n. We can assume that the shift vectors of the $R_i \in P'$ are zero modulo T^n. It follows that $(W,\vec{v})(R_i,0)=(WR_i,\vec{v})$ for $R_i \in P'$. If the symmetry support U_1 of WR_1 is orthogonal to $\vec{v}$ then $(WR_1,\vec{v})$ has a symmetry support $\vec{p}_1+U_1$ which remains pointwise fixed (cf. section 5.5). Let $M:=\{(WR_1,\vec{v}),\dots,(WR_k,\vec{v})\}$ be a set of symmetry operations which leave at least one common point fixed then M is the generating set of a finite subgroup of G equivalent to $P''<P$. The group P" may define another regular splitting of $O(P,x_0)$. From this immediately the following theorem is deduced:

<u>Theorem 8.6:</u> If a crystallographic point group P has a symmetry operation $S \in P$ which leaves only the origin fixed then any space group G derived from an arithmetic crystal class (P,T^n) has a finite subgroup equivalent to the point group generated by S.

Proof: By assumption the symmetry support of S is 0-dimensional hence, it is orthogonal to every shift vector $\vec{s}$. It follows that $(S,\vec{s})$ leaves a point fixed. Thus $(S,\vec{s})$ generates a point group of finite order.

<u>Theorem 8.7:</u> From a cyclic $\mathbb{Z}$-irreducible crystallographic point group P with arithmetic crystal class (P,T^n) arises only the symmorphic space group type.

Proof: Let P',C' be a regular splitting of $O(P,x_0)$. Since P is cyclic and P',C' have only the identity operation in common it follows that there exists at least one coset $P'W_i$ having a symmetry operation S which generates P. Since S is $\mathbb{Z}$-irreducible it leaves only the origin fixed. Therefore, the symmetry support of S is orthogonal to every vector $\vec{v}$. It follows that for any shift vector $\vec{s}$, the symmetry operation $(S,\vec{s})$ generates a finite subgroup of G equivalent to P. Thus G is equivalent to the symmorphic space group of (P,T^n).

According to Plesken each $\mathbb{Z}$-irreducible crystallographic point group in dimensions up to n=21 only determines one arithmetic crystal class (Ples4).

The following theorem is due to Schwarzenberger (Schw1), and Michel and Mozrzymas (Mich1):

<u>Theorem 8.8:</u> If the crystallographic point group P has the central inversion -I then for any space group G derived from (P,T^n) the shift vectors $\vec{s}$ can be chosen such that $\vec{s}=\vec{t}/2$, $\vec{t}\in T^n$.

Proof: The central inversion -I leaves only the origin fixed. Hence, the origin of G can be chosen such that $(-I,0)\in G$. For any symmetry operation $(S,\vec{s})\in G$ it follows that

$$(-I,0)(S,\vec{s})(-I,0)(S,\vec{s})^{-1} = (I,2\vec{s})$$

is a translation which proves theorem 8.8.

Example 8.1: There are 13 arithmetic crystal classes in two dimensional space E^2 (cf. section 7.4). Considering these arithmetic crystal classes in turn the space group types in E^2 can be derived. Space groups in E^2 are called plane groups. For the plane group types the nomenclature of Hermann-Mauguin is used (for details the reader is referred to the International Tables Vol. A, Hahn1).

1. From the arithmetic crystal classes (C_1,T^2) and (C_2,T^2) only the symmorphic plane groups p1 and p2 respectively arise because by theorem 8.6 every plane group derived from these classes contains points of symmetry C_1 and C_2 respectively hence, they are equivalent to the symmorphic plane groups.

2. From the arithmetic crystal classes (C_3,T^2), (C_4,T^2), and (C_6,T^2) by theorem 8.7 only the symmorphic plane group types p3, p4, and p6 respectively arise.

3. Two arithmetic crystal classes belong to the crystal class C_s, viz.: (C_s,T_p^2) and (C_s,T_c^2). We consider the regular splitting $P':=C_1$, $C':=C_s$. We determine $N:=N_{/A}(\Sigma')\cap N_{/A}(F')$. The quotient T'/T_p^2 has translations $\vec{s}_1:=0$, $\vec{s}_2:=\varepsilon\vec{b}$ and $\vec{s}_3:=\frac{1}{2}\vec{a}$ (cf. Figure 7.3). The shift vector $\vec{s}_1$ gives the symmorphic plane group type pm. Since P' is of index 2 in P it follows that $\vec{s}_2:=\frac{1}{2}\vec{b}$ is a possible shift vector giving the plane group type pg. The shift vector $\vec{s}_3$ is orthogonal to the mirror line therefore, an equivalent mirror line is generated thus the derived space group is equivalent to the zero-solution.

   Similarly, T'/T_c^2 contains the possible shift vectors $\vec{s}_1:=0$, $\vec{s}_2:=\frac{1}{2}(\vec{a}_1-\vec{a}_2)$. The shift vector $\vec{s}_1$ gives the symmorphic space group type cm. Because $\vec{s}_2$ is orthogonal to the mirror line m it follows that an equivalent mirror line is generated hence, the space group thus generated is equivalent to the zero-solution.

4. Two arithmetic crystal classes belong to C_{2v} viz.: (C_{2v},T_p^2) and (C_{2v},T_c^2) (cf. Figure 7.2). The crystallographic point group C_{2v} contains the central inversion $-I$ hence, by theorem 8.6 every space group derived from C_{2v} contains a subgroup equivalent to C_2. Thus we only have to consider the regular splitting $P':=C_2$, $C':=C_s$. Both groups P' and C' are normal in P. We calculate $N:=N_{/A}(\Sigma')\cap N_{/A}(F')$. The qoutient group T'/T_p^2 has the shift vectors $\vec{s}_1:=0$, $\vec{s}_2:=\frac{1}{2}\vec{a}$, $\vec{s}_3:=\frac{1}{2}\vec{b}$, and $\vec{s}_4:=\frac{1}{2}(\vec{a}+\vec{b})$. The shift vector $\vec{s}_1$ gives the symmorphic plane group type p2mm. The shift vectors $\vec{s}_2$ and $\vec{s}_3$ are equivalent under the normalizer of C_{2v} therefore, they give rise to the same space group type p2mg. The shift vector $\vec{s}_4$ gives rise to the plane group type p2gg.

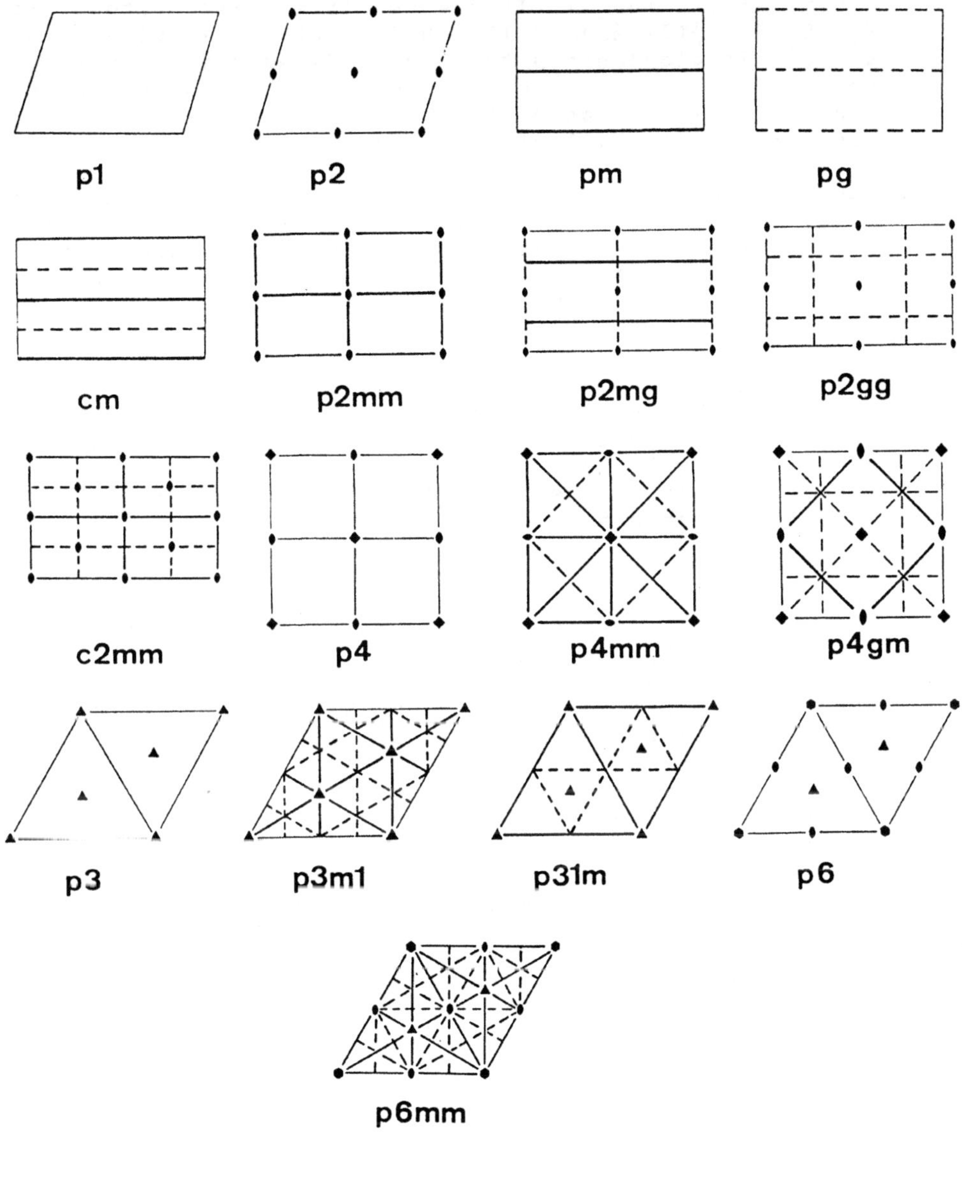

Figure 8.1. Graphical representation of fragments of the symmetry scaffoldings belonging to represenatatives of the 17 plane group types

Similarly, for T'/T_c^2 the possible shift vectors are $\vec{s}_1:=0$, $\vec{s}_2:=\frac{1}{2}(\vec{a}_1-\vec{a}_2)$. The shift vector $\vec{s}_1$ gives the symmorphic plane group type c2mm. Because $\vec{s}_2$ is orthogonal to the mirror line m it follows that an equivalent mirror line is generated hence, the space group thus generated is equivalent to the zero-solution.

5. Two arithmetic crystal classes belong to C_{3v}, viz.: (C_{3v},T_α^2) and (C_{3v},T_β^2) (cf. Figure 7.4). By theorem 8.6, every plane group derived from C_{3v} contains a subgroup equivalent to C_3. Therefore, we only have to consider the regular splitting $P':=C_3$, $C':=C_s$. Only C_3 is normal in C_{3v}. We determine $N:=N_{/A}(\Sigma')$. The quotient group T'/T_α^2 has shift vectors $\vec{s}_1:=0$, $\vec{s}_2:=\frac{1}{3}\vec{a}_1+\frac{2}{3}\vec{a}_2$, $\vec{s}_3:=\frac{2}{3}\vec{a}_1+\frac{1}{3}\vec{a}_2$. The shift vector $\vec{s}_1$ gives the symmorphic plane group p31m. The shift vectors $\vec{s}_2$ and $\vec{s}_3$ are equivalent under the normalizer of C_{3v}. None of them is compatible with a symmetry operation of order 2. Therefore, no further solution results. Similarly, for (C_{3v},T_β^2) we obtain only the symmorphic plane group type p3m1.

6. We consider the arithmetic crystal class (C_{4v},T^2). By theorem 8.6, every plane group derived from C_{4v} contains a subgroup equivalent to C_4. Therefore, only the regular splitting $P':=C_4$, $C':=C_s$ has to be considered. Only C_4 is normal in C_{4v}. We determine $N:=N_{/A}(\Sigma')$. The quotient group T'/T^2 has the shift vectors $\vec{s}_1:=0$, $\vec{s}_2:=\frac{1}{2}(\vec{a}_1+\vec{a}_2)$. The shift vector $\vec{s}_1$ gives the symmorphic plane group type p4mm. The shift vector $\vec{s}_2$ gives the plane group type p4gm.

7. We consider the arithmetic crystal class (C_{6v},T^2). By theorem 8.6, every plane group derived from C_{6v} contains a subgroup equivalent to C_6. Therefore, only the regular splitting $P':=C_6$, $C':=C_s$ has to be considered. Only C_6 is normal in C_{6v}. We determine $N:=N_{/A}(\Sigma')$. The quotient group T'/T^2 has only the shift vector $\vec{s}_1:=0$. Thus only the symmorphic plane group type p6mm results.

The three-dimensional space groups were determined by Schönflies (Schö2) and Fedorov (Fedo3). The four-dimensional space groups were determined by Brown et al. (Brow1). The results of these calculations are summerized in the following theorem:

Theorem 8.9: There exist 2 in 1-dimensional, 17 in 2-dimensional, 219+11 in 3-dimensional and 4783+112 space group types in 4-dimensional space.

In spaces of higher dimensions only partial results are known. Schwarzenberger has investigated the "primitiv orthogonal" and the "primitiv tetragonal" groups using a graphical method (Schw1).

8.3. Normalizers of symmetry groups

For many theoretical investigations of symmetry groups their normalizers play an important role.

Definition 8.7: For a group Γ with subgroup G the normalizer of G in Γ, $N_\Gamma(G)$, is the group of all $\varphi\in\Gamma$ under which G is conjugated to itself,

$$N_\Gamma(G) := \{\varphi \mid \varphi\in\Gamma,\ \varphi G\varphi^{-1}=G\}.$$

The name "normalizer" indicates that the subgroup G is normal in $N_\Gamma(G)$. For most applications the normalizer is taken in the Euclidean group $\mathbb{E}(n)$ of all isometries, or in the affine group $\mathbb{A}(n)$ of all affine transformations.

In section 7.4 use was made of the normalizer $N_O(P)$ of a point group P in the orthogonal group O(n) of all rotations. Necessarily the normalizer $N_O(P)$ maps the symmetry scaffolding of P onto itself. However, the stabilizer of the symmetry scaffolding of P in O(n) may be of higher order than $N_O(P)$ because if U is an invariant subspace of P then U is also an invariant subspace of $N_O(P)$. For each $\varphi\in N_O(P)$, $\varphi P=P\varphi$ implies that φU is an inavriant subspace of P. If the invariant subspaces of P are uniquely determined it follows that $\varphi U=U$. For cyclic groups U need not belong to the symmetry scaffolding of P. The normalizers $N_O(P)$ of the three-dimensional crystallographic point groups were determined by Galiulin (Gali1). The normalizers of the four-dimensional crystallographic point groups were determined by Brown (Brow1).

In section 8.2 the affine normalizers of space groups were used. Since space groups in E^n contain n linearly independent translations it follows that they have no

Table 8.1. The normalizers $N_0(P)$ of the three-dimensional crystallographic point groups

N (P)	crystal classes
K_h	C_1, C_i
$D_{\infty h}$	$C_s, C_2, C_{2h}, C_3, C_{3i}, C_4, S_4, C_{4h}, C_{3h}, C_6, C_{6h}$
D_{12h}	D_6, C_{6v}, D_{6h}
D_{8h}	D_4, C_{4v}, D_{4h}
D_{6h}	$D_3, C_{3v}, D_{3d}, D_{3h}$
D_{4h}	C_{2v}, D_{2d}
O_h	$D_2, D_{2h}, T, T_h, T_d, O, O_h$

proper invariant subspace. Therefore, the normalizer of a space group G in the affine group $\mathbb{A}(n)$ is the group of all affine mappings $\varphi \in \mathbb{A}(n)$ which map the symmetry scaffolding of G onto itself. This gives a simple way to determine the normalizer of a space group.

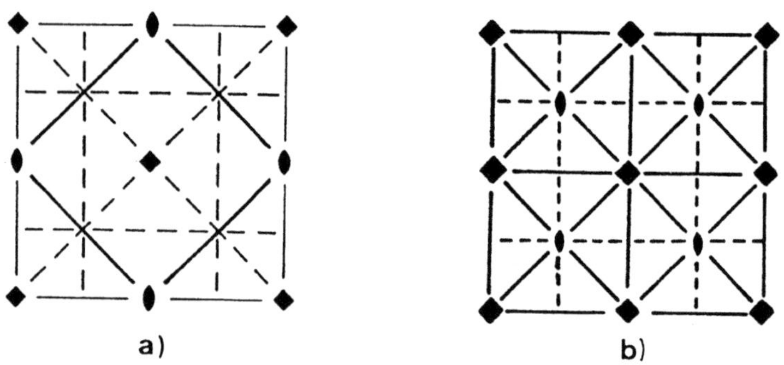

Figure 8.2. Fragments of the symmetry scaffoldings of a) the plane group p4gm, and b) the normalizer $N_{\mathbb{E}}$(p4gm)

Example 8.2: We consider the plane group type p4gm. A fragment of the symmetry scaffolding of p4gm is shown in Figure 8.2 a). The normalizer

$$N_{\mathbb{E}}(p4gm)=p4mm\ [\tfrac{1}{2}(\vec{a}_1-\vec{a}_2),\tfrac{1}{2}(\vec{a}_1+\vec{a}_2)]$$

contains all symmetry operations which map the symmetry scaffolding of p4gm onto itself. It has an additional translation $\frac{1}{2}(\vec{a}_1+\vec{a}_2)$. A lattice basis of N_{IE}(p4gm) is given by the lattice vectors

$$\tfrac{1}{2}(\vec{a}_1-\vec{a}_2),\ \tfrac{1}{2}(\vec{a}_1+\vec{a}_2).$$

A fragment of the symmetry scaffolding of the normalizer N_{IE}(p4gm) is shown in Figure 8.2 b).

By definition 8.7 the normalizer $N_\Gamma(G)$ is defined for an individual group G only. In general the normalizers, belonging to an equivalence class of space groups, do not belong to one equivalence class. This is only true if the equivalence class of G and the normalizer of G are taken in the same group Γ. Thus for the space groups G_i belonging to the same equivalence class their Euclidean normalizers $N_{IE}(G_i)$ do not necessarily belong to the same equivalence class. Following Schwarzenberger (Schw1) there exists in each type of space groups a minimal group G such that the lattice of each space group equivalent to G has at least the symmetry of the lattice of G. Now we can take out of each type of space groups such a minimal space group and determine its normalizer. The normalizers $N_{IE}(G)$ of the three-dimensional minimal space groups were determined by Hirshfeld (Hirs1). The affine normalizers of space groups were determined by Burzlaff, Zimmermann, and Billiet (Burz2, Bill1), and by Gubler (Gubl1).

Table 8.2. The normalizers $N_{IE}(G)$ of the two-dimensional minimal plane groups

$N_{IE}(G)$	basis of $N_{IE}(G)$	minimal plane groups G
$z^2 2$	$\varepsilon\vec{a}\times\varepsilon\vec{b}$	p1
p2	$\frac{1}{2}\vec{a}\times\frac{1}{2}\vec{b}$	p2
$z^1 2mm$	$\frac{1}{2}\vec{a}\times\varepsilon\vec{b}$	pm, pg, cm
p2mm	$\frac{1}{2}\vec{a}\times\frac{1}{2}\vec{b}$	p2mm, p2mg, p2gg, c2mm
p4mm	$\frac{1}{2}(\vec{a}-\vec{b})\times\frac{1}{2}(\vec{a}+\vec{b})$	p4, p4mm, p4gm
p6mm	$\vec{a}\times\vec{b}$	p31m, p6, p6mm
	$\frac{1}{3}(\vec{a}-\vec{b})\times\frac{1}{3}(\vec{a}+2\vec{b})$	p3, p3m1

Table 8.3. The normalizers $N_{IE}(G)$ of the three-dimensional minimal space groups

$N_{IE}(G)$	basis of $N_{IE}(G)$	minimal space groups G
$P\bar{1}$	$\frac{1}{2}\vec{a}\times\frac{1}{2}\vec{b}\times\frac{1}{2}\vec{c}$	$P\bar{1}$
$Z^3\bar{1}$	$\varepsilon\vec{a}\times\varepsilon\vec{b}\times\varepsilon\vec{c}$	$P1$
$P2/m$	$\frac{1}{2}\vec{a}\times\frac{1}{2}\vec{b}\times\frac{1}{2}\vec{c}$	$P2/m, P2_1/m, C2/m, P2/c, P2_1/c, C2/c$
$Z^1 2/m$	$\frac{1}{2}\vec{a}\times\varepsilon\vec{b}\times\frac{1}{2}\vec{c}$	$P2, P2_1, C2$
$Z^2 2/m$	$\varepsilon\vec{a}\times\frac{1}{2}\vec{b}\times\varepsilon\vec{c}$	Pm, Pc, Cm, Cc
$Pmmm$	$\frac{1}{2}\vec{a}\times\frac{1}{2}\vec{b}\times\frac{1}{2}\vec{c}$	$P222, P222_1, P2_12_12, P2_12_12_1, C222, C222_1, I222, I2_12_12_1, Pmmm, Pnnn, Pccm, Pban, Pmma, Pnna, Pmna, Pcca, Pbam, Pccn, Pbcm, Pnnm, Pmmn, Pbcn, Pbca, Pnma, Cmcm, Cmca, Cmmm, Cccm, Cmma, Ccca, Fmmm, Immm, Ibam, Ibca, Imma$
$Pnnn$	$\frac{1}{2}\vec{a}\times\frac{1}{2}\vec{b}\times\frac{1}{2}\vec{c}$	$Fddd$
$Immm$	$\frac{1}{2}\vec{a}\times\frac{1}{2}\vec{b}\times\frac{1}{2}\vec{c}$	$F222$
Z^1mmm	$\frac{1}{2}\vec{a}\times\frac{1}{2}\vec{b}\times\varepsilon\vec{c}$	$Pmm2, Pmc2_1, Pcc2, Pma2, Pca2_1, Pnc2, Pmn2_1, Pba2, Pna2_1, Pnn2, Cmm2, Cmc2_1, Ccc2, Amm2, Abm2, Ama2, Aba2, Fmm2, Imm2, Iba2, Ima2$
Z^1ban	$\frac{1}{2}\vec{a}\times\frac{1}{2}\vec{b}\times\varepsilon\vec{c}$	$Fdd2$
$P4_222$	$\frac{1}{2}(\vec{a}-\vec{b})\times\frac{1}{2}(\vec{a}+\vec{b})\times\frac{1}{2}\vec{c}$	$P4_122, P4_12_12, P4_322, P4_32_12$
Z^1422	$\frac{1}{2}(\vec{a}-\vec{b})\times\frac{1}{2}(\vec{a}+\vec{b})\times\varepsilon\vec{c}$	$P4_1, P4_3$
$P4/mmm$	$\frac{1}{2}(\vec{a}-\vec{b})\times\frac{1}{2}(\vec{a}+\vec{b})\times\frac{1}{2}\vec{c}$	$P\bar{4}, P4/m, P4_2/m, P4/n, P4_2/n, I4/m, P422, P42_12, P4_222, P4_22_12, I422, P\bar{4}2m. P\bar{4}2c, P\bar{4}2_1m, P\bar{4}2_1c, P\bar{4}m2, P\bar{4}c2, P\bar{4}b2, P\bar{4}n2, I\bar{4}2m, P4/mmm, P4/mcc, P4/nbm, P4/nnc, P4/mbm, P4/mnc, P4/nmm, P4/ncc, P4_2/mmc, P4_2/mcm, P4_2/nbc, P4_2/nnm, P4_2/mbc, P4_2/mnm, P4_2/nmc, P4_2/ncm, I4/mmm, I4/mcm$
$P4_2/nnm$	$\frac{1}{2}(\vec{a}-\vec{b})\times\frac{1}{2}(\vec{a}+\vec{b})\times\frac{1}{2}\vec{c}$	$I4_1/a, I4_122, I\bar{4}2d, I4_1/amd, I4_1/acd$
$I4/mmm$	$\frac{1}{2}(\vec{a}-\vec{b})\times\frac{1}{2}(\vec{a}+\vec{b})\times\frac{1}{2}\vec{c}$	$I\bar{4}, I\bar{4}m2, I\bar{4}c2$
Z^14/mmm	$\frac{1}{2}(\vec{a}-\vec{b})\times\frac{1}{2}(\vec{a}+\vec{b})\times\varepsilon\vec{c}$	$P4, P4_2, I4, P4mm, P4bm, P4_2cm, P4_2nm, P4cc, P4nc, P4_2mc, P4_2bc, I4mm, I4cm$
Z^14/nbm	$\frac{1}{2}(\vec{a}-\vec{b})\times\frac{1}{2}(\vec{a}+\vec{b})\times\varepsilon\vec{c}$	$I4_1, I4_1md, I4_1cd$
$R\bar{3}m$	$-\vec{b}\times(\vec{a}+\vec{b})\times\frac{1}{2}\vec{c}$	$R\bar{3}, R32, R\bar{3}m, R\bar{3}c$
$Z^1\bar{3}1m$	$\frac{1}{3}(\vec{a}-\vec{b})\times\frac{1}{3}(\vec{a}+2\vec{b})\times\varepsilon\vec{c}$	$R3, R3m, R3c$
$P6_222$	$\vec{a}\times\vec{b}\times\frac{1}{2}\vec{c}$	$P3_121, P6_122, P6_422$
	$\frac{1}{3}(\vec{a}-\vec{b})\times\frac{1}{3}(\vec{a}+2\vec{b})\times\frac{1}{2}\vec{c}$	$P3_112$
$P6_422$	$\vec{a}\times\vec{b}\times\frac{1}{2}\vec{c}$	$P3_221, P6_522, P6_222$
	$\frac{1}{3}(\vec{a}-\vec{b})\times\frac{1}{3}(\vec{a}+2\vec{b})\times\frac{1}{2}\vec{c}$	$P3_212$

Table 8.3 (continued)

$N_{IE}(G)$	basis of $N_{IE}(G)$	minimal space groups G
Z^1622	$\vec{a}\times\vec{b}\times\varepsilon\vec{c}$	$P6_1, P6_5, P6_2, P6_4$
	$\frac{1}{3}(\vec{a}-\vec{b})\times\frac{1}{3}(\vec{a}+2\vec{b})\times\varepsilon\vec{c}$	$P3_1, P3_2$
P6/mmm	$\vec{a}\times\vec{b}\times\frac{1}{2}\vec{c}$	$P\bar{3}, P321, P\bar{3}1m, P\bar{3}1c, P\bar{3}m1, P\bar{3}c1, P6/m, P6_3/m, P622, P6_322, P\bar{6}2m, P\bar{6}2c, P6/mmm, P6/mcc, P6_3/mcm, P6_3/mmc$
	$\frac{1}{3}(\vec{a}-\vec{b})\times\frac{1}{3}(\vec{a}+2\vec{b})\times\frac{1}{2}\vec{c}$	$P312, P\bar{6}, P\bar{6}m2, P\bar{6}c2$
Z^16/mmm	$\vec{a}\times\vec{b}\times\varepsilon\vec{c}$	$P31m, P31c, P6, P6_3, P6mm, P6cc, P6_3cm, P6_3mc$
	$\frac{1}{3}(\vec{a}-\vec{b})\times\frac{1}{3}(\vec{a}+2\vec{b})\times\varepsilon\vec{c}$	$P3, P3m1, P3c1$
Ia3	$\vec{a}\times\vec{b}\times\vec{c}$	Pa3
$I4_132$	$\vec{a}\times\vec{b}\times\vec{c}$	$P4_332, P4_132$
$Pm\bar{3}m$	$\frac{1}{2}\vec{a}\times\frac{1}{2}\vec{b}\times\frac{1}{2}\vec{c}$	$Fm\bar{3}, F432, Fm\bar{3}m, Fm\bar{3}c$
$Pn\bar{3}m$	$\frac{1}{2}\vec{a}\times\frac{1}{2}\vec{b}\times\frac{1}{2}\vec{c}$	$Fd\bar{3}, F4_132, Fd\bar{3}m, Fd\bar{3}c$
$Im\bar{3}m$	$\vec{a}\times\vec{b}\times\vec{c}$	$P23, I23, Pm\bar{3}, Pn\bar{3}, Im\bar{3}, P432, P4_232, I432, P\bar{4}3m, I\bar{4}3m, P\bar{4}3n, Pm\bar{3}m, Pn\bar{3}n, Pm\bar{3}n, Pn\bar{3}m, Im\bar{3}m$
	$\frac{1}{2}\vec{a}\times\frac{1}{2}\vec{b}\times\frac{1}{2}\vec{c}$	$F23, F\bar{4}3m, F\bar{4}3c$
$Ia\bar{3}d$	$\vec{a}\times\vec{b}\times\vec{c}$	$P2_13, I2_13, Ia\bar{3}, I4_132, I\bar{4}3d, Ia\bar{3}d$

In general, the normalizer of a space group is not a space group. It may contain infinitely small translations of length ε in d linearly independent directions. The lattice thus generated is denoted by z^d or Z^d for the normalizers of the two- and three-dimensional space groups respectively. Normalizers of space groups in the affine group $\mathbb{A}$ may also contain point groups of infinite order.

8.4. Subgroups of space groups

For many applications the subgroups G' of a space group G in E^n are required. We consider for the moment only subgroups of finite index that is G' is again a space group in E^n (for subgroups of infinite index see section 8.7). Let $P' \leq P$ be the crystallographic point groups and $T' \leq T$ be the lattices of G' and G respectively. Following Hermann (Herm2), three kinds of subgroups are distinguished:

- translation equivalent subgroups with $T'=T$,
- class equivalent subgroups with $P'=P$,
- general subgroups with $T'<T$ and $P'<P$.

The following two theorems are due to Hermann (Herm2):

Theorem 8.10: For each space group G there exists a translation equivalent subgroup H such that any subgroup G' of G is a class equivalent subgroup of H,

$$G' \leq H \leq G.$$

Proof: The theorem follows from the fact that the subgroup H is uniquely determined by G' and T.

A subgroup $G'<G$ is called maximal in G if there exists no other subgroup G'' such that

$$G' < G'' < G.$$

Theorem 8.11: Every maximal subgroup G' of a space group G is either translation equivalent or class equivalent.

Proof: Because G' is maximal either $G'<H=G$ then G' is class equivalent or else $G'=H<G$ then G' is translation equivalent.

8.4.1. Derivation of subgroups

We now describe a procedure to determine the subgroups of a space group. Following Senechal (Sene2), a space group G is written as a union of cosets with respect to T,

$$G = T \cup T(S_2,\vec{s}_2) \cup \ldots \cup T(S_s,\vec{s}_s),$$

and T is written as a union of cosets with respect to T',

$$T = T' \cup T'\vec{v}_2 \cup \ldots \cup T'\vec{v}_v.$$

We are free to choose the shift vectors $\vec{s}_i$ modulo T and the lattice vectors $\vec{v}_i$ modulo T'. Combining both coset decompositions yields

$$\begin{array}{llll} G = T' & \cup T'(S_2,\vec{s}_2) & \cup \ldots & \cup T'(S_s,\vec{s}_s) \\ \cup T'\vec{v}_2 & \cup T'(S_2,\vec{s}_2+\vec{v}_2) & \cup \ldots & \cup T'(S_s,\vec{s}_s+\vec{v}_2) \\ \vdots & & & \vdots \\ \cup T'\vec{v}_v & \cup T'(S_2,\vec{s}_2+\vec{v}_v) & \cup \ldots & \cup T'(S_s,\vec{s}_s+\vec{v}_v). \end{array}$$

A subgroup G' having crystallographic point group P' and lattice T' is the union of some of these cosets under the conditions that the group is closed under coset multiplication and that P' maps the lattice T' onto itself.

The invariance of T' under P' requires that for every $S_i \in P'$ and each basis vector $\vec{b}_j \in T'$, $j=1,\ldots,n$,

$$S_i\vec{b}_j = \vec{t} \in T'.$$

Let h' be the order of P'. The closure of the coset multiplication requires that

$$T'(S_i,\vec{s}_i+\vec{v}_\lambda)\ T'(S_j,\vec{s}_j+\vec{v}_\mu) = T'(S_k,\vec{s}_k+\vec{v}_\nu),$$

where $S_k=S_iS_j$, $i,j=1,\ldots,h'$. It follows that

$$T'(S_i\vec{s}_j+S_i\vec{v}_\mu+\vec{s}_i+\vec{v}_\lambda) = T'(S_k,\vec{s}_k+\vec{v}_\nu).$$

The translation parts on both sides of this equation have to be equal up to a translation $\vec{t}_{ij} \in T'$. Thus we get a Frobenius subgroup congruence,

$$S_i\vec{s}_j+S_i\vec{v}_\mu+\vec{s}_i+\vec{v}_\lambda-\vec{s}_k-\vec{v}_\nu = 0 \bmod T'.$$

A similar congruence is set up for each pair $S_i,S_j \in P'$. A set of shift vectors $\vec{s}_1+\vec{v}_\alpha,\ldots,\vec{s}_h+\vec{v}_\omega$ which fulfils all h'^2 congruences gives a possible subgroup G'<G.

From a group theoretic point of view two subgroups G' and G" are equivalent if there exists an affine mapping $\varphi \in N_{/A}(G)$ such that

$$G'' = \varphi G' \varphi^{-1}.$$

However, if we consider the action of a space group G on a regular point system X then two subgroups $G', G'' < G$ which are equivalent under $N_{\mathbb{A}}(G)$ may act differently on X. In this case only the equivalence under the group G is allowed. Thus, depending on the problem we are treating, different definitions of the equivalence of subgroups are required. Therefore, we may consider the equivalence under the normalizer of G in $\mathbb{A}(n)$, $\mathbb{A}^+(n)$, $\mathbb{E}(n)$, $\mathbb{E}^+(n)$, G, or G^+. For physical problems usually the equivalence under G is required.

For a given P' and T' the equivalence condition for $\varphi := (R, \vec{r}) \in N_\Gamma(G)$ becomes

$$\varphi(S_i, \vec{s}_i + \vec{v}_\lambda)\varphi^{-1} = (S_j, \vec{s}_j + \vec{v}_\mu) \bmod T', \quad (i=1,\ldots,h'),$$

where $S_j = RS_iR^{-1}$. Let L be the lattice of $N_\Gamma(G)$ with lattice basis $\vec{c}_1,\ldots,\vec{c}_n$. Each $\varphi \in N_\Gamma(G)$ is written as

$$\varphi := (R, \vec{r} + m_1\vec{c}_1 + \ldots + m_n\vec{c}_n), \quad (m_i \in \mathbb{Z}).$$

From the above congruence it follows that

$$(\vec{r} - S_i\vec{r}) - (\vec{s}_i - R\vec{s}_i) - (\vec{v}_\mu - R\vec{v}_\lambda) + m_1(\vec{c}_1 - S_i\vec{c}_1) + \ldots + m_n(\vec{c}_n - S_i\vec{c}_n)$$
$$= 0 \bmod T'.$$

The vectors $\vec{t}_{1i} := \vec{c}_1 - S_i\vec{c}_1, \ldots, \vec{t}_{ni} := \vec{c}_n - S_i\vec{c}_n$, $i=1,\ldots,h'$, generate a lattice T", called the weight lattice of L with respect to P'. As the above congruence has to be taken modulo T' we only need consider vectors $\vec{t}$ in the quotient group T"/T'. Thus the equivalence of two subgroups G' and G" can be verified. Special care has to be taken if $N_\Gamma(G)$ has a point group of infinite order or if it contains infinitely small translations.

The number of subgroups G' of a space group G is infinite. However, these may be grouped into a finite number of series, each containing only subgroups of the same space group type, which are obtained by increasing the index of the sublattice T' in T. For a given crystallographic point group P'<P this requires the determination of all lattices T'<T which are invariant under P'. Let $\vec{b}_1,\ldots,\vec{b}_n$ be a lattice basis of T. A lattice basis $\vec{b}'_1,\ldots,\vec{b}'_n$ of T' is obtained trough a transformation with

transformation matrix $A:=(a_{ij})$ having integral coefficients a_{ij},

$$\begin{pmatrix} \vec{b}'_1 \\ \cdot \\ \cdot \\ \cdot \\ \vec{b}'_n \end{pmatrix} = \begin{pmatrix} a_{11} & \dots & a_{1n} \\ \cdot & & \cdot \\ \cdot & & \cdot \\ \cdot & & \cdot \\ a_{n1} & \dots & a_{nn} \end{pmatrix} \begin{pmatrix} \vec{b}_1 \\ \cdot \\ \cdot \\ \cdot \\ \vec{b}_n \end{pmatrix}.$$

The index of T' in T is equal to det(A). The number of lattice bases belonging to a lattice T' is infinite. They all fall into one equivalence class (cf. section 3.2). We have to restrict the coefficients a_{ij} such that for each sublattice T' only one lattice basis is obtained. This can be achieved by considering the transformations into the different Bravais lattices belonging to P'. Thereby, also unconventional Bravais types of lattices occur. Finally among all lattices T' thus obtained we have to determine those which are not equivalent under $N_T(G)$.

In what follows we give the necessary transformations to obtain the required sublattices of the three-dimensional space groups. It is assumed that the standard setting of the space groups is used as given in the International Tables, Vol. A. For conventionally centered lattices having basis vectors $\vec{a}_1, \vec{a}_2, \vec{a}_3$ first a transformation with transformation matrix $B:=(b_{ij})$ is applied to obtain a lattice basis $\vec{b}_1, \vec{b}_2, \vec{b}_3$,

$$\begin{pmatrix} \vec{b}_1 \\ \vec{b}_2 \\ \vec{b}_3 \end{pmatrix} = \begin{pmatrix} b_{11} & b_{12} & b_{13} \\ b_{21} & b_{22} & b_{23} \\ b_{31} & b_{32} & b_{33} \end{pmatrix} \begin{pmatrix} \vec{a}_1 \\ \vec{a}_2 \\ \vec{a}_3 \end{pmatrix}.$$

The following transformations are used:

A-lattice	C-lattice	I-lattice	F-lattice
$\begin{pmatrix} 1 & 0 & 0 \\ 0 & \frac{1}{2} & -\frac{1}{2} \\ 0 & \frac{1}{2} & \frac{1}{2} \end{pmatrix}$	$\begin{pmatrix} \frac{1}{2} & -\frac{1}{2} & 0 \\ \frac{1}{2} & \frac{1}{2} & 0 \\ 0 & 0 & 1 \end{pmatrix}$	$\begin{pmatrix} 1 & 0 & 0 \\ 0 & 1 & 0 \\ \frac{1}{2} & \frac{1}{2} & \frac{1}{2} \end{pmatrix}$	$\begin{pmatrix} \frac{1}{2} & \frac{1}{2} & 0 \\ 0 & \frac{1}{2} & \frac{1}{2} \\ \frac{1}{2} & 0 & \frac{1}{2} \end{pmatrix}$

In Table 8.4 the transformation matrices A_i are given, referred to a lattice basis $\vec{b}_1, \vec{b}_2, \vec{b}_3$. By theorem 7.1, every lattice remains invariant under the central inversion −I. Thus we have to consider only the crystallographic point groups L generated by L:={P',−I}. The groups L are called Laue groups (after the physicist Max von Laue). The corresponding equivalence classes are called Laue classes. For each Laue class L the transformations among all Bravais types of lattices, invariant under L, are given in Table 8.4. If P' and P belong to the same Laue class then Table 8.4 can directly be applied. However, if P' belongs to another Laue class than P then the lattice basis $\vec{b}_1, \vec{b}_2, \vec{b}_3$ of G first is transformed into a lattice basis $\vec{c}_1, \vec{c}_2, \vec{c}_3$ corresponding to the standard setting for P'. The transformation matrix is $W := (w_{ij})$. The transformations A_i are applied to the lattice basis $\vec{c}_1, \vec{c}_2, \vec{c}_3$. Thus we get the transformations $A'_i = A_i W$. In this case however, sublattices T' which are equivalent under P occur. For centered lattices we obtain the transformations, referred to the conventional basis, through $A''_i = A_i W B$.

Example 8.3: In three-dimensional space we consider the space group $Fm\bar{3}m$ with point group O_h-$m\bar{3}m$ and face centered lattice F. We ask for all non-equivalent subgroups belonging to $P' := C_4$-4. A lattice basis of $Fm\bar{3}m$ is given by

$$\begin{aligned} \vec{b}_1 &= \tfrac{1}{2}\vec{a}_1 + \tfrac{1}{2}\vec{a}_2 + 0\vec{a}_3 \\ \vec{b}_2 &= 0\vec{a}_1 + \tfrac{1}{2}\vec{a}_2 + \tfrac{1}{2}\vec{a}_3 \\ \vec{b}_3 &= \tfrac{1}{2}\vec{a}_1 + 0\vec{a}_2 + \tfrac{1}{2}\vec{a}_3 . \end{aligned}$$

Suppose P' is the fourfold axis along the direction $-\vec{b}_1 + \vec{b}_2 + \vec{b}_3$ (conventional $\vec{a}_3$-direction). For P' the F-lattice corresponds to a standard tetragonal I-lattice. The transformation with transformation matrix W becomes,

$$\begin{vmatrix} \vec{c}_1 \\ \vec{c}_2 \\ \vec{c}_3 \end{vmatrix} = \begin{vmatrix} 1 & 0 & 0 \\ 0 & 1 & -1 \\ 0 & 1 & 0 \end{vmatrix} \begin{vmatrix} \vec{b}_1 \\ \vec{b}_2 \\ \vec{b}_3 \end{vmatrix} .$$

The Laue class of C_4 is C_{4h}-4/m. In Table 8.4 there are two kinds of transformations A_i, viz.: A_1: I→I and A_2: I→P. We calculate $A'_i = A_i W$, i=1,2.

$$A'_1 = \begin{vmatrix} a & d & -d \\ -d & a & -a \\ \tfrac{1}{2}(a-c-d) & \tfrac{1}{2}(a+c+d) & \tfrac{1}{2}(-a+c-d) \end{vmatrix} \quad \begin{array}{l} a-c+d=2n;\ d<a \\ \text{index } a^2c+d^2c \end{array}$$

$$A'_2 = \begin{vmatrix} a & d & -d \\ -d & a & -a \\ -c & c & c \end{vmatrix} \quad \begin{array}{l} d<a \\ \text{index } 2c(a^2+d^2). \end{array}$$

Since P' belongs to a different Laue class than P we further require that d<a in order to eliminate equivalent lattices. We determine $A''_i = A_i W B$, i=1,2, which give the transformations, referred to the conventional basis of $Fm\overline{3}m$,

$$A''_1 = \begin{vmatrix} \frac{1}{2}(a-d) & \frac{1}{2}(a+d) & 0 \\ -\frac{1}{2}(a+d) & \frac{1}{2}(a-d) & 0 \\ -\frac{1}{2}d & \frac{1}{2}a & \frac{1}{2}c \end{vmatrix} \quad a-c+d=2n;\ d<a$$

$$A''_2 = \begin{vmatrix} \frac{1}{2}(a-d) & \frac{1}{2}(a+d) & 0 \\ -\frac{1}{2}(a+d) & \frac{1}{2}(a-d) & 0 \\ 0 & 0 & c \end{vmatrix} \quad d<a.$$

If a,c,d run through all permitted integers then all non-equivalent lattices, invariant under P', are obtained. Thus we get the required series:

A''_1 for c=2n: $I4$, $I4_1$
for c=2n+1: $I4$

A''_2 for c=2n: $P4$, $P4_1$, $P4_2$
for c=2n+1: $P4$, $P4_2$

Table 8.4. Transformation matrices for the generation of sublattices T' for each Laue class L in E^3 $(a,d,c \in \mathbb{N},\ d,e,f \in \mathbb{N}_0)$

1. Laue class C_i – $\bar{1}$ (triclinic)

P→P $\begin{vmatrix} a & 0 & 0 \\ d & b & 0 \\ e & f & c \end{vmatrix}$ d,e<a; f<b; index abc

2. Laue class C_{2h} – 2/m (monoclinic unique axis b)

P→P $\begin{vmatrix} a & 0 & 0 \\ 0 & b & 0 \\ d & 0 & c \end{vmatrix}$ d<a; index abc

P→C $\begin{vmatrix} a & -b & 0 \\ a & b & 0 \\ d & 0 & c \end{vmatrix}$ d<a; index 2abc

P→A $\begin{vmatrix} a & 0 & d \\ 0 & b & -c \\ 0 & b & c \end{vmatrix}$ d<c; index 2abc

P→I $\begin{vmatrix} a & 0 & 0 \\ \frac{1}{2}(a+d) & b & \frac{1}{2}c \\ d & 0 & c \end{vmatrix}$ a+d=2n; c=2n; a≥2; d<a; index abc

C→C $\begin{vmatrix} \frac{1}{2}(a+b) & \frac{1}{2}(a-b) & 0 \\ \frac{1}{2}(a-b) & \frac{1}{2}(a+b) & 0 \\ d & d & c \end{vmatrix}$ a+b=2n; d<a; index abc

C→P $\begin{vmatrix} a & a & 0 \\ -b & b & 0 \\ d & d & c \end{vmatrix}$ d<a; index 2abc

C→A $\begin{vmatrix} a & a & d \\ -b & b & -c \\ -b & b & c \end{vmatrix}$ d<a; index 4abc

C→I $\begin{vmatrix} a & a & 0 \\ \frac{1}{2}(a-2b+d) & \frac{1}{2}(a+2b+d) & \frac{1}{2}c \\ d & d & c \end{vmatrix}$ a+d=2n; c=2n; a≥2; d<a; index 2abc

3. Laue class D_{2h} – mmm (orthorhombic)

P→P $\begin{vmatrix} a & 0 & 0 \\ 0 & b & 0 \\ 0 & 0 & c \end{vmatrix}$ index abc

P→A $\begin{vmatrix} a & 0 & 0 \\ 0 & b & -c \\ 0 & b & c \end{vmatrix}$ index 2abc

P→B $\begin{vmatrix} a & 0 & -c \\ 0 & b & 0 \\ a & 0 & c \end{vmatrix}$ index 2abc

P→C $\begin{vmatrix} a & -b & 0 \\ a & b & 0 \\ 0 & 0 & c \end{vmatrix}$ index 2abc

P→F $\begin{vmatrix} a & b & 0 \\ 0 & b & c \\ a & 0 & c \end{vmatrix}$ index 2abc

P→I $\begin{vmatrix} 2a & 0 & 0 \\ 0 & 2b & 0 \\ a & b & c \end{vmatrix}$ index 4abc

Table 8.4. (continued)

$$A\to A\quad \begin{vmatrix} a & 0 & 0 \\ 0 & \frac{1}{2}(b+c) & \frac{1}{2}(b-c) \\ 0 & \frac{1}{2}(b-c) & \frac{1}{2}(b+c) \end{vmatrix} \begin{matrix} b+c=2n \\ \text{index } abc \\ \ \end{matrix}$$

$$A\to P\quad \begin{vmatrix} a & 0 & 0 \\ 0 & b & b \\ 0 & -c & c \end{vmatrix} \ \text{index } 2abc$$

$$A\to B\quad \begin{vmatrix} a & c & -c \\ 0 & b & b \\ a & -c & c \end{vmatrix} \ \text{index } 4abc$$

$$A\to C\quad \begin{vmatrix} a & -b & -b \\ a & b & b \\ 0 & -c & c \end{vmatrix} \ \text{index } 4abc$$

$$A\to F\quad \begin{vmatrix} a & b & b \\ 0 & b-c & b+c \\ a & -c & c \end{vmatrix} \ \text{index } 4abc$$

$$A\to I\quad \begin{vmatrix} 2a & 0 & 0 \\ 0 & b & b \\ a & \frac{1}{2}(b-c) & \frac{1}{2}(b+c) \end{vmatrix} \begin{matrix} b+c=2n \\ \text{index } 2abc \\ \ \end{matrix}$$

$$C\to C\quad \begin{vmatrix} \frac{1}{2}(a+b) & \frac{1}{2}(a-b) & 0 \\ \frac{1}{2}(a-b) & \frac{1}{2}(a+b) & 0 \\ 0 & 0 & c \end{vmatrix} \begin{matrix} a+b=2n \\ \text{index } abc \\ \ \end{matrix}$$

$$C\to P\quad \begin{vmatrix} a & a & 0 \\ -b & b & 0 \\ 0 & 0 & c \end{vmatrix} \ \text{index } 2abc$$

$$C\to A\quad \begin{vmatrix} a & a & 0 \\ -b & b & -c \\ -b & b & c \end{vmatrix} \ \text{index } 4abc$$

$$C\to B\quad \begin{vmatrix} a & a & -c \\ -b & b & 0 \\ a & a & c \end{vmatrix} \ \text{index } 4abc$$

$$C\to F\quad \begin{vmatrix} a-b & a+b & 0 \\ -b & b & c \\ a & a & c \end{vmatrix} \ \text{index } 4abc$$

$$C\to I\quad \begin{vmatrix} a & a & 0 \\ -b & b & 0 \\ \frac{1}{2}(a-b) & \frac{1}{2}(a+b) & c \end{vmatrix} \begin{matrix} a+b=2n \\ \text{index } 2abc \\ \ \end{matrix}$$

$$I\to I\quad \begin{vmatrix} a & 0 & 0 \\ 0 & b & 0 \\ \frac{1}{2}(a-c) & \frac{1}{2}(b-c) & c \end{vmatrix} \begin{matrix} a+c=2n;\ b+c=2n \\ \text{index } abc \\ \ \end{matrix}$$

$$I\to P\quad \begin{vmatrix} a & 0 & 0 \\ 0 & b & 0 \\ -c & -c & 2c \end{vmatrix} \ \text{index } 2abc$$

$$I\to A\quad \begin{vmatrix} a & 0 & 0 \\ 0 & b+c & -2c \\ -c & b-c & 2c \end{vmatrix} \ \text{index } 4abc$$

$$I\to B\quad \begin{vmatrix} a+c & c & -2c \\ 0 & b & 0 \\ a-c & -c & 2c \end{vmatrix} \ \text{index } 4abc$$

$$I\to C\quad \begin{vmatrix} a & -b & 0 \\ a & b & 0 \\ -c & -c & 2c \end{vmatrix} \ \text{index } 4abc$$

$$I\to F\quad \begin{vmatrix} a & b & 0 \\ -c & b-c & 2c \\ a-c & -c & 2c \end{vmatrix} \ \text{index } 4abc$$

Table 8.4. (continued)

$$F\to F \quad \begin{vmatrix} \frac{1}{2}(a+b) & \frac{1}{2}(b-a) & \frac{1}{2}(a-b) \\ \frac{1}{2}(b-c) & \frac{1}{2}(b+c) & \frac{1}{2}(c-b) \\ \frac{1}{2}(a-c) & \frac{1}{2}(c-a) & \frac{1}{2}(a+c) \end{vmatrix}$$

$a+b=2n$; $b+c=2n$; $a+c=2n$; index abc

$$F\to P \quad \begin{vmatrix} a & -a & a \\ b & b & -b \\ -c & c & c \end{vmatrix}$$

index $4abc$

$$F\to A \quad \begin{vmatrix} a & -a & a \\ \frac{1}{2}(b+c) & \frac{1}{2}(b-c) & -\frac{1}{2}(b+c) \\ \frac{1}{2}(b-c) & \frac{1}{2}(b+c) & -\frac{1}{2}(b-c) \end{vmatrix}$$

$b+c=2n$; index $2abc$

$$F\to B \quad \begin{vmatrix} \frac{1}{2}(a+c) & -\frac{1}{2}(a+c) & \frac{1}{2}(a-c) \\ b & b & -b \\ \frac{1}{2}(a-c) & -\frac{1}{2}(a-c) & \frac{1}{2}(a+c) \end{vmatrix}$$

$a+c=2n$; index $2abc$

$$F\to C \quad \begin{vmatrix} \frac{1}{2}(a-b) & -\frac{1}{2}(a+b) & \frac{1}{2}(a+b) \\ \frac{1}{2}(a+b) & -\frac{1}{2}(a-b) & \frac{1}{2}(a-b) \\ -c & c & c \end{vmatrix}$$

$a+b=2n$; index $2abc$

$$F\to I \quad \begin{vmatrix} a & -a & a \\ b & b & -b \\ \frac{1}{2}(a+b-c) & \frac{1}{2}(b+c-a) & \frac{1}{2}(a+c-b) \end{vmatrix}$$

$a+b+c=2n$; index $2abc$

4. Laue class C_{4h} – 4/m (tetragonal)

$$P\to P \quad \begin{vmatrix} a & d & 0 \\ -d & a & 0 \\ 0 & 0 & c \end{vmatrix}$$

index a^2c+d^2c

$$P\to I \quad \begin{vmatrix} a & d & 0 \\ -d & a & 0 \\ \frac{1}{2}(a-d) & \frac{1}{2}(a+d) & c \end{vmatrix}$$

$a+d=2n$; index a^2c+d^2c

$$I\to I \quad \begin{vmatrix} a & d & 0 \\ -d & a & 0 \\ \frac{1}{2}(a-c-d) & \frac{1}{2}(a-c+d) & c \end{vmatrix}$$

$a-c+d=2n$; index a^2c+d^2c

$$I\to P \quad \begin{vmatrix} a & d & 0 \\ -d & a & 0 \\ -c & -c & 2c \end{vmatrix}$$

index $2c(a^2+d^2)$

5. Laue class D_{4h} – 4/mmm (tetragonal)

The same as for Laue class 4/m with $d=0$ or $d=a$ only.

Table 8.4. (continued)

6. Laue class C_{3i} – $\bar{3}$ (trigonal)

$$P\to P \begin{vmatrix} a & d & 0 \\ -d & a-d & 0 \\ 0 & 0 & c \end{vmatrix} \text{ index } (a^2-ad+d^2)c$$

$$P\to R \begin{vmatrix} a & d & a \\ -d & a-d & c \\ d-a & -a & c \end{vmatrix} \text{ index } 3c(a^2-ad+d^2)$$

$$R\to R \begin{vmatrix} \frac{1}{3}(2a-d+c) & \frac{1}{3}(2d-a+c) & \frac{1}{3}(c-a-d) \\ \frac{1}{3}(c-d-a) & \frac{1}{3}(2a-d+c) & \frac{1}{3}(2d-a+c) \\ \frac{1}{3}(2d-a+c) & \frac{1}{3}(c-d-a) & \frac{1}{3}(2a-d+c) \end{vmatrix} \begin{array}{l} 2a-d+c=3n \\ \text{index } (a^2-ad+d^2)c \end{array}$$

$$R\to P \begin{vmatrix} a & d-a & -d \\ -d & a & d-a \\ c & c & c \end{vmatrix} \text{ index } 3c(a^2-ad+d^2)$$

7. Laue class D_{3d} – $\bar{3}m$ (trigonal)

The same as for Laue class $\bar{3}$ with d=0 or d=a/2 only.

8. Laue class C_{6h} – 6/m (hexagonal)

$$P\to P \begin{vmatrix} a & d & 0 \\ -d & a-d & 0 \\ 0 & 0 & c \end{vmatrix} \begin{array}{l} d<a \\ \text{index } c(a^2-ad+d^2) \end{array}$$

9. Laue class D_{6h} – 6/mmm (hexagonal)

The same as for Laue class 6/m with d=0 or d=a/2 only.

10. Laue class T_h – $m\bar{3}$ (cubic)

$$P\to P \begin{vmatrix} a & 0 & 0 \\ 0 & a & 0 \\ 0 & 0 & a \end{vmatrix} \text{ index } a^3 \qquad P\to I \begin{vmatrix} a & a & 0 \\ 0 & a & a \\ a & 0 & a \end{vmatrix} \text{ index } 2a^3$$

$$P\to I \begin{vmatrix} 2a & 0 & 0 \\ 0 & 2a & 0 \\ a & a & a \end{vmatrix} \text{ index } 4a^3$$

$$F\to F \begin{vmatrix} a & 0 & 0 \\ 0 & a & 0 \\ 0 & 0 & a \end{vmatrix} \text{ index } a^3 \qquad F\to P \begin{vmatrix} a & -a & a \\ a & a & -a \\ -a & a & a \end{vmatrix} \text{ index } 4a^3$$

Table 8.4. (continued)

$F \to I \begin{vmatrix} 2a & -2a & 2a \\ 2a & 2a & -2a \\ a & a & a \end{vmatrix}$ index $16a^3$ $\quad I \to I \begin{vmatrix} a & 0 & 0 \\ 0 & a & 0 \\ 0 & 0 & a \end{vmatrix}$ index a^3

$I \to P \begin{vmatrix} a & 0 & 0 \\ 0 & a & 0 \\ -a & -a & 2a \end{vmatrix}$ index $2a^3$ $\quad I \to F \begin{vmatrix} a & a & 0 \\ -a & 0 & 2a \\ 0 & -a & 2a \end{vmatrix}$ index $4a^3$

11. Laue class O_h - $m\overline{3}m$ (cubic)

The same as for Laue class $m\overline{3}$.

8.4.2. Topological characterization of subgroups

The determination of subgroups requires a procedure which allows a quick identification of the space group type. Such a procedure should be independent on the choice of lattice basis and of the origin. Therefore, only geometric properties of a space group are considered which remain invariant under affine transformations. Thus we get a topological characterization of a space group.

Let G be a space group with point group P and lattice T. Properties of G which remain invariant under affine transformations are the order of P and the occurence or absence of rotations or screw rotations in G. We identify each minimax symmetry operation $(S_i,\vec{s}_i)\in G$ through its double weighted symmetry support $\vec{p}_i+U_i$ (cf. section 8.1). Thereby, it is sufficient to consider only those symmetry supports $\vec{p}_i+U_i$ which have a non-empty intersection with a fundamental region F of G. Since F is bounded we obtain a finite set of symmetry supports. The occuring subspaces U_i are completely determined through the crystallographic point group P. Besides the symmetry supports we have to determine their mutual intersections. Thereby, for dimensions $n>2$, also the intersections of the symmetry supports of screw rotations or glide reflections are needed. Moreover, if a subspace U_i has dimension $d>1$ and screw vector $\vec{t}_i$ then the orientatin of $\vec{t}_i$, relative to the symmetry scaffolding of G, may be important. A similar case occurs if two symmetry operations, having different screw vectors $\vec{t}_1$ and $\vec{t}_2$ respectively, belong to two cosets of the same subspace U_i. Thus a set of distinct topological properties of a space group can be determined which uniquely identifies the space group type.

Example 8.4: We consider the 17 plane group types. A possible set of topological properties of the plane groups is given in the headline of Table 8.4. For each plane group type the occurence or absence of a property is indicated by 1 or 0 in the corresponding column. Thus every plane group type is uniquely characterized through a binary bit string of length 9.

Table 8.4. Topological characterization of the 17 plane groups

	2	3	4	6	m	g	2mm	3m	4mm
p1	0	0	0	0	0	0	0	0	0
p2	1	0	0	0	0	0	0	0	0
pm	0	0	0	0	1	0	0	0	0
pg	0	0	0	0	0	1	0	0	0
cm	0	0	0	0	1	1	0	0	0
p2mm	0	0	0	0	1	0	1	0	0
p2mg	1	0	0	0	1	1	0	0	0
p2gg	1	0	0	0	0	1	0	0	0
c2mm	1	0	0	0	1	1	1	0	0
p4	1	0	1	0	0	0	0	0	0
p4mm	0	0	0	0	1	1	1	0	1
p4gm	0	0	1	0	1	1	1	0	0
p3	0	1	0	0	0	0	0	0	0
p3m1	0	0	0	0	1	1	0	1	0
p31m	0	1	0	0	1	1	0	1	0
p6	1	1	0	1	0	0	0	0	0
p6mm	0	0	0	0	1	1	1	1	0

The International Tables, Vol. A lists only the maximal subgroups of the three-dimensional space groups. The present procedure allows to determine all subgroups of a space group in form of a finite number of series. A computer program has been worked out using the above method to determine the subgroups of space groups (Enge7).

8.4.3. Normal subgroups

Of particular interest are the normal subgroups of space groups. The following theorem is due to Senechal (Sene3).

<u>Theorem 8.12:</u> Let G' be a subgroup of a space group G with crystallographic point groups P',P and lattices T',T respectively. G' is normal in G if and only if

1. P' is normal in P,
2. $S_i\vec{b}'_j \in T'$ for every $S_i \in P$ and every basis vector $\vec{b}'_j \in T'$,
3. $\vec{b}_j - R_i\vec{b}_j = 0 \bmod T'$ for every $R_i \in P'$ and every basis vector $\vec{b}_j \in T$,
4. $S_i\vec{r}_j - R_j\vec{s}_i + \vec{s}_i - \vec{r}_k = 0 \bmod T'$.

Proof: G' is normal in G if for every $(S_i, \vec{s}_i + \vec{t}) \in G$ and every $(R_j, \vec{r}_j + \vec{t}') \in G'$ we have that

$$(S_i, \vec{s}_i + \vec{t})(R_j, \vec{r}_j + \vec{t}')(S_i, \vec{s}_i + \vec{t})^{-1}$$

$$= (R_k, -R_k\vec{s}_i - R_k\vec{t} + S_i\vec{r}_j + S_i\vec{t}' + \vec{s}_i + \vec{t})$$
$$= (R_k, \vec{r}_k) \bmod T',$$

with $R_k = S_iR_jS_i^{-1}$. This requires that

1. $R_k \in P'$,
2. for $R_j = I$ and every $\vec{t}' \in T'$, $S_i\vec{t}' \in T'$, for every $S_i \in P$,
3. for $S_i = I$ and every $\vec{t} \in T$, $\vec{t} - R_j\vec{t} = 0 \bmod T'$, for every $R_j \in P'$,
4. because of conditions 1 to 3 we have

$$S_i\vec{r}_j - R_j\vec{s}_i + \vec{s}_i - \vec{r}_k = 0 \bmod T'.$$

Let $\vec{b}'_1, \ldots, \vec{b}'_n$ be a lattice basis of T'. Every lattice vector $\vec{t}' \in T'$ is uniquely written as $\vec{t}' = m'_1\vec{b}'_1 + \ldots + m'_n\vec{b}'_n$, $m'_i \in \mathbb{Z}$. Condition 2 requires that for each $\vec{t}' \in T'$ also $S_i\vec{t}' \in T'$, for every $S_i \in P$. This is already fulfilled if for each basis vector $\vec{b}'_j \in T'$ also $S_i\vec{b}'_j \in T'$, for every $S_i \in P$. Similarly, let $\vec{b}_1, \ldots, \vec{b}_n$ be a lattice basis of T. Condition 3 is fulfilled if for each basis vector $\vec{b}_i \in T$ also $\vec{b}_i - R_j\vec{b}_i \in T'$, for every $R_j \in P'$.

Complete lists of normal subgroups of the two- and three-dimensional space groups were calculated by the author. For the two-dimensional space groups the images G/G' were determined by Senechal (Sene3).

8.5. Crystallographic orbits

Orbits of symmetry groups were defined in section 8.1. Particularly an orbit generated by a space group is called a crystallographic orbit because of its discrete nature and its periodicity. In section 7.5 we investigated orbits of crystallographic point groups. These were classified according to their adjoint crystal forms. In a similar way we now classify crystallographic orbits.

8.5.1. Classifications of crystallographic orbits

In n-dimensional Euclidean space E^n we consider a crystallographic orbit $O(G,x_0)$. The space group G is called the generating space group and the point x_0 is called a representing point of the crystallographic orbit. Let $St(x_0)<G$ be the stabilizer of the point x_0 in G. Equivalent points $x_i \in O(G,x_0)$, $x_i:=(S_i,\vec{s}_i)x_0$, have conjugate stabilizers in G,

$$St(x_i)=(S_i,\vec{s}_i)St(x_0)(S_i,\vec{s}_i)^{-1}, \quad (S_i,\vec{s}_i)\in G.$$

In the International Tables, Vol. A, the crystallographic orbits of a space group G are classified according to the stabilizer of the representing point in the following way:

Definition 8.8: The set of all points y which have conjugate stabilizers St(y) in G is called a Wyckoff position, if conjugation is taken in G. It is called a Wyckoff set, if conjugation is taken in the affine normalizer $N_{/A}(G)$.

The above definition can be visulized if we consider the symmetry support $\vec{p}_0+U_0$ of $St(x_0)$. The subspace U_0 has dimension d, $0\le d\le n$. We identify a Wyckoff position or a Wyckoff set with the set of all symmetry supports $\vec{p}_i+U_i$ which are equivalent to $\vec{p}_0+U_0$ under G or $N_{/A}(G)$ respectively.

Wyckoff positions or Wyckoff sets only classify crystallographic orbits of one space group. We now consider ordered pairs (G,W) of a space group G and a Wyckoff set W belonging to G.

<u>Definition 8.9:</u> Two ordered pairs (G_1,W_1) and (G_2,W_2) are called equivalent and belong to the same type of Wyckoff sets if there is an affine mapping $\varphi\in\mathbb{A}(n)$ which maps G_1 onto G_2 and simultaneously maps W_1 onto W_2,

$$G_2 = \varphi G_1\varphi^{-1}, \quad W_2 = \varphi W_1.$$

Because of physical reasons it may be useful to consider only affine mappings φ^+ of the special affine group $\mathbb{A}^+(n)$ which conserve the chirality of a crystallographic orbit.

In three-dimensional space the types of Wyckoff sets were determined by Koch and Fischer (Koch2) by inspection of the space group Table. The results are summerized in the following theorem:

<u>Theorem 8.13:</u> There exist 3 in 1-dimensional, 51 in 2-dimensional and 1128+29 types of Wyckoff sets in 3-dimensional space.

Each crystallographic orbit corresponds to a regular point system $X\subset\mathbb{E}^n$. The types of Wyckoff sets do not classify regular point systems because a certain regular point system X may be generated from points belonging to different types of Wyckoff sets. A classification of the set of all regular point systems is obtained by considering the stabilizer of X in the Euclidean group $\mathbb{E}(n)$ of all isometries. By definition 1.3 the stabilizer of X in $\mathbb{E}(n)$ is a space group called the eigensymmetry space group of X. The stabilizer G_s of a crystallographic orbit $O(G,x_0)$ may be a supergroup of the generating space group G, $G_s\geq G$. A crystallographic orbit is called characteristic or non-characteristic with respect to G if $G_s=G$ or $G_s>G$ respectively. Each regular point system X can be assigned to a stabilizer G_s and a unique Wyckoff set W_s in G_s.

<u>Definition 8.10:</u> Two regular point systems X_1 and X_2 with stabilizers and Wyckoff sets (G_{s1},W_{s1}) and (G_{s2},W_{s2}) respectively are called equivalent and belong to the same orbit type if there exists an affine mapping $\varphi\in\mathbb{A}(n)$ which maps G_{s1} onto G_{s2} and simultaneously maps W_{s1} onto W_{s2},

$$G_{s2} = \varphi G_{s1}\varphi^{-1}, \quad W_{s2} = \varphi W_{s1}.$$

A still finer classification of crystallographic orbits is obtained if we consider their topological properties. We define the topology of a regular point system X through its Dirichlet domain partition (cf. sections 2.1 to 2.3, and 9.2). In complete analogy to definition 7.5 this allows us to classify regular point systems in topological types of orbits. Results on Dirichlet domain partitions will be given in section 9.2.

8.5.2. Determination of crystallographic orbits

In order to determine the non-characteristic orbits of a space group G an analogous procedure, as described in section 7.5, can be applied. Let $O(G,x_0)$ be a non-characteristic orbit with stabilizer $G_s>G$ of index k. Suppose the point x_0 has stabilizer $St(x_0)$ in G and $St_s(x_0)$ in G_s. The ratio of the orders of $St_s(x_0)$ and of $St(x_0)$ has to be equal to the index k,

$$\frac{|St_s(x_0)|}{|St(x_0)|} = k. \qquad (8.5)$$

because otherwise the stabilizer G_s would generate more points than are contained in $O(G,x_0)$. Thus the special Wyckoff positions W_s in the stabilizer G_s which fulfil equation 8.5 are solutions for non-characteristic orbits of G. Among the infinite number of supergroups of G we have to find the stabilizers G_s of $O(G,x_0)$. This can be done in the following way: If there exists another space group G' between G and G_s, $G<G'<G_s$, then $O(G',x_0)$ must also have the stabilizer G_s since $O(G_s,x_0)$ does not split into two orbits when going from G_s to G. Hence, if no special Wyckoff position exists in G' which satisfies equation 8.5 then G' and any supergroup G">G' cannot be a stabilizer of $O(G,x_0)$. Therefore, all solutions for non-characteristic orbits of G are already contained in the minimal supergroups G' of G. From theorem 8.11 it follows that every minimal supergroup G' is either class equivalent or translation equivalent to G. If G' is translation equivalent then the index k' of G' in G is less or equal to p, where p is the largest prime factor in the order h of the crystallographic point group P of G. If G' is class equivalent then $k'\le h/|St(x_0)|$. Thus only a finite number of supergroups have to be considered in order to find all non-characteristic orbits of G. The range of supergroups can be further restricted if we compare the affine normalizers of G and G'. We determine the intersection

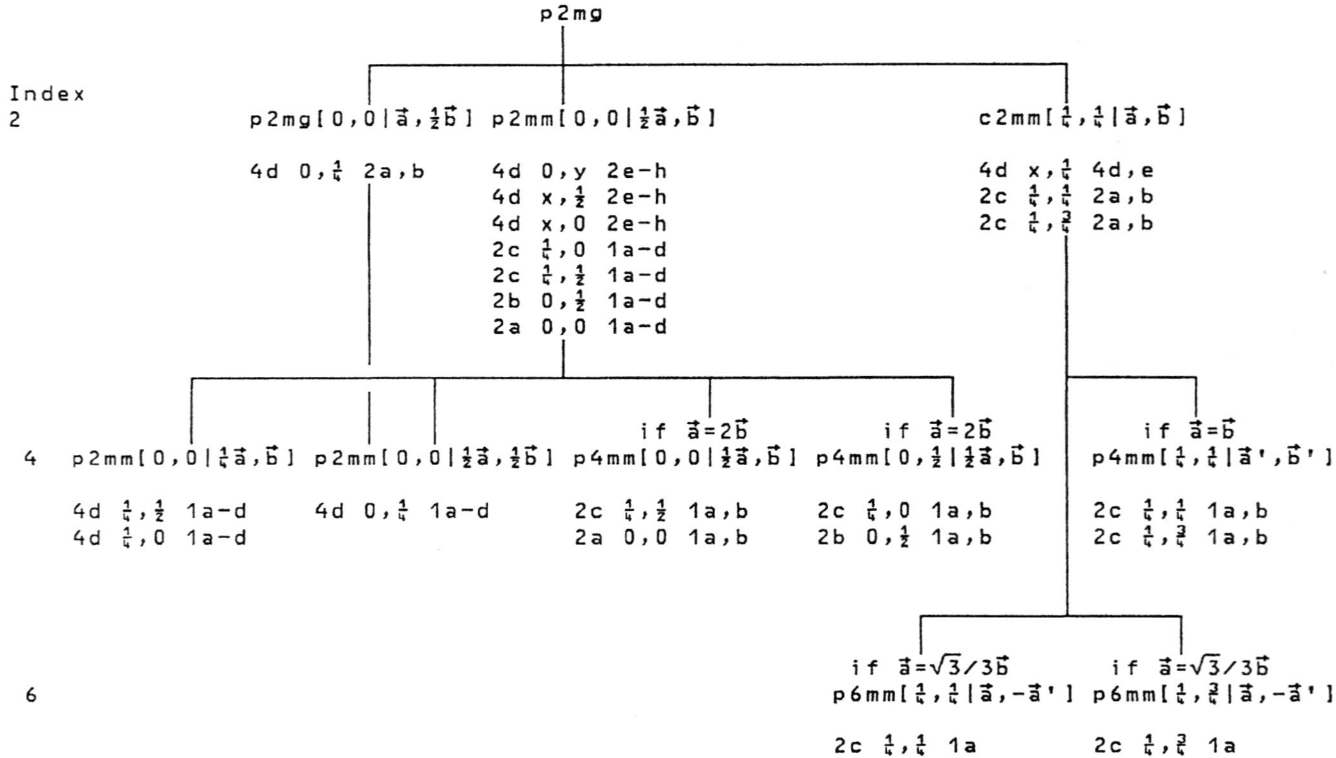

Figure 8.3. The non-characteristic orbits of the plane group p2mg
$\vec{a}' := \frac{1}{2}(\vec{a}-\vec{b})$, $\vec{b}' := \frac{1}{2}(\vec{a}+\vec{b})$

$$N := N_{/A}(G) \cap N_{/A}(G').$$

If $N=N_{/A}(G)$ then we find in G' all solutions which are equivalent under $N_{/A}(G)$. However, if $N<N_{/A}(G)$ with index v then we have to consider also supergroups conjugated under $N_{/A}(G)$. We determine the coset decomposition of $N_{/A}(G)$ with respect to N,

$$N_{/A}(G) = N \cup \rho_2 N \cup \ldots \cup \rho_v N.$$

The required conjugated supergroups are given by $G'_j := \rho_j G' \rho_j^{-1}$, $j=2,\ldots,v$. If several supergroups G'_i, $i=1,\ldots,m$, have a common supergroup G" then we find all solutions in G" by intersecting the solutions in G'_i, $i=1,\ldots,m$. Finally the stabilizer G_s of X is the highest supergroup which still contains a solution for X. Usually there are several paths from the space group G to the stabilizer G_s. It is sufficient to determine one.

Example 8.4: We consider the plane group p2mg. The standard setting of the International Table, Vol. A, is assumed. The required group-supergroup relations which lead to the non-characteristic orbits of p2mg are shown in Figure 8.3. The supergroups are given with the Hermann-Mauguin symbol followed, between brackets, by the origin and the basis vectors, referred to the original plane group p2mg. Beneath each supergroup are listed the solutions for non-characteristic orbits according to the Wyckoff positions of p2mg. Each set of solutions is followed by the symbol of the Wyckoff set in the supergroup. The affine normalizer of p2mg is $p2mm[\frac{1}{2}\vec{a},\frac{1}{2}\vec{b}]$. We determine the intersection

$$N := N_{/A}(p2mg) \cap N_{/A}(p4mm[\tfrac{1}{2}\vec{a},\vec{b}]) = p2mm[\tfrac{1}{2}\vec{a},\vec{b}].$$

Since $N<N_{/A}(p2mg)$ of index 2 we determine the coset decopmosition of $N_{/A}(2pmg)$ with respect to N and obtain as a coset representative $\rho_2 := \frac{1}{2}\vec{b}$. Similarly, we determine

$$N := N_{/A}(p2mg) \cap N_{/A}(p6mm[\vec{a},-\vec{a}']) = c2mm[\vec{a},\vec{b}],$$

where $\vec{a}' = \frac{1}{2}(\vec{a}-\vec{b})$. N is of index 2 in $N_{/A}(p2mg)$ and we obtain $\rho_2 := \frac{1}{2}\vec{b}$ as a coset representative.

In the plane E^2 only 13 out of 17 plane-group types occur as stabilizers of regular point systems. These correspond to Sohncke's 13 construction types (Sohn2). The

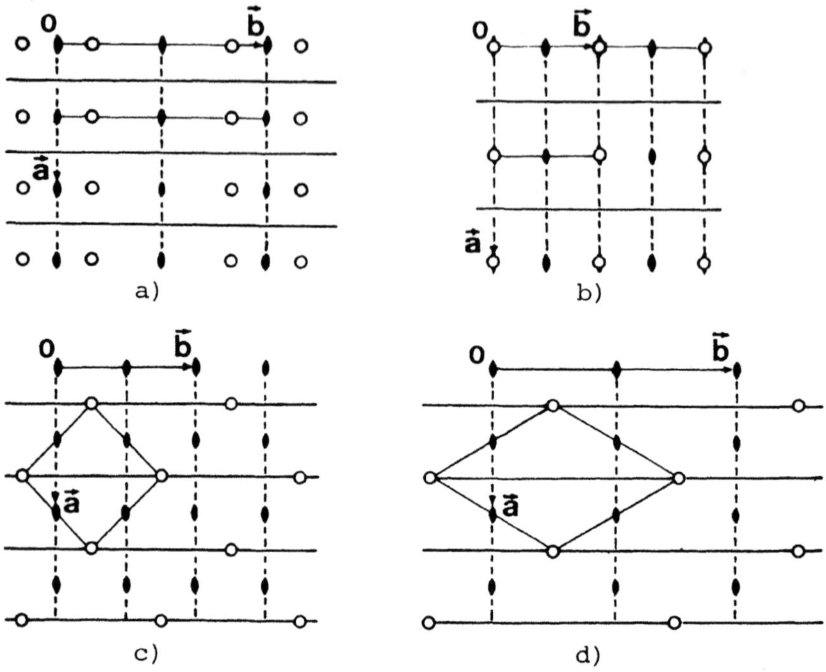

Figure 8.4. Examples of non-characteristic orbits in the plane group p2mg. a) 4d 0,y, b) 2a 0,0 ($\vec{a}=2\vec{b}$), c) 2c $\frac{1}{4},\frac{1}{4}$ ($\vec{a}=\vec{b}$), d) 2c $\frac{1}{4},\frac{1}{4}$ ($\vec{a}=\sqrt{3}/3\vec{b}$)

non-characteristic orbits, given in Table 8.5, are specialized cases of these construction types. The non-characteristic orbits were independently determined again by Lawrenson and Wondratschek (Lawr1) for the plane-group types having minimal symmetry of the lattice, that is, no accidental higher lattice symmetry is accounted for. In three-dimensional space E^3 only 166+11 out of 219+11 space-group types occur as stabilizers of regular point systems. The non-characteristic orbits of the space-group types, having minimal symmetry of the lattice, were determined by Engel, Matsumoto, Steinmann, and Wondratschek (Enge4). The results of these investigations are summerized in the following theorem:

Table 8.5. Sohncke's 13 construction types and the possible stabilizers of regular point systems in the plane E^2

Plane group type	construction type	possible stabilizers
p2	XI	p2, p2mm, p2mg, c2mm, p4mm, p6mm
p2mm	X	p2mm, p4mm
p2mg	XII	p2mg, p2mm, c2mm, p4mm, p6mm
p2gg	XIII	p2gg, p2mg, p2mm, c2mm, p4mm, p4gm, p6mm
c2mm	VI	c2mm, p2mm, p4mm, p6mm
p4	IV	p4, p4mm, p4gm
p4mm	VII	p4mm
p4gm	V	p4gm, p4mm
p3	III	p3, p3m1, p31m, p6mm
p3m1	IX	p3m1, p6mm
p31m	II	p31m, p6mm
p6	I	p6, p6mm
p6mm	VIII	p6mm

The underscored plane groups are documented by Figures in Sohncke's paper.

<u>Theorem 8.14:</u> There exist 2 in 1-dimensional, 30 in 2-dimensional, and 402+25 orbit types in 3-dimensional space.

8.5.3. Determination of congruent crystallographic orbits

For a given crystallographic orbit $O(G,x_0)$ we now determine the set of all orbits $O(G,y_0)$, $y_0 \notin O(G,x_0)$, which are congruent to $O(G,x_0)$, that is, there exists an isometry $\varphi \in \mathbb{E}(n)$ such that $O(G,y_0) = \varphi O(G,x_0)$. If $O(G,x_0)$ is characteristic in G, $G_s = G$, then we obtain all congruent crystallographic orbits through the application of the Euclidean normalizer $N_{\mathbb{E}}(G)$. We determine the coset decomposition

$$N_{\mathbb{E}}(G) = G \cup \rho_2 G \cup \ldots \cup \rho_k G.$$

Since $N_{\mathbb{E}}(G)$ contains all isometries which map the symmetry scaffolding of G onto itself it follows that the coset representatives ρ_i, $i=2,\ldots,k$, give all congruent crystallographic orbits $O(G,y_{0i})$ with $y_{0i} := \rho_i x_0$. However, if $O(G,x_0)$ is non-characteristic with stabilizer $G_{s1} > G$ then additional congruent crystallographic orbits may occur. We have to consider all supergroups G_{si}, $i=2,\ldots,m$, which are not equivalent to G_{s1} under the Euclidean normalizer $N_{\mathbb{E}}(G)$. Let $\varphi_i \in \mathbb{E}(n)$ be an isometry which maps G_{s1} onto G_{si},

$$G_{si} := \varphi_i G_{s1} \varphi_i^{-1}.$$

Thereby, the point x_0 is mapped onto $x_{0i} := \varphi_i x_0$. In order to find all solutions in each supergroup G_{si}, $i=1,\ldots,m$, we compare the Euclidean normalizers $N_{\mathbb{E}}(G)$ and $N_{\mathbb{E}}(G_{si})$. We determine the intersection

$$N_i := N_{\mathbb{E}}(G) \cap N_{\mathbb{E}}(G_{si}).$$

If $N_i \leq N_{\mathbb{E}}(G)$ then further solutions are obtained as follows: we determine the coset decomposition

$$N_{\mathbb{E}}(G_{si}) = \rho_{i1} N_i \cup \rho_{i2} N_i \cup \ldots \cup \rho_{iv} N_i,$$

where ρ_{i1} is the identity operation. For $x_{0ij} := \rho_{ij} x_{0i}$, $j=1,\ldots,v$, the crystallographic orbit $O(G, x_{0ij})$ is congruent to $O(G, x_0)$ exactly if equation 8.5 is fulfilled, that is, if $|St(x_{0ij})| = |St(x_0)|$. Again all points equivalent to x_{0ij} under $N_{\mathbb{E}}(G)$ generate congruent crystallographic orbits.

8.5.4. Principal points of space groups

A crystallographic orbit $O(G, x_0)$ is called a point lattice if all of its points are translationally equivalent under some space group $G_s \geq G$. The point lattice is called a real point lattice if its points are equivalent under the translations of the generating point group G. Otherwise it is called an apparent point lattice. It was shown by Nowacki (Nowa4) that in three-dimensional space groups, except of those of type $P4_132$ ($P4_332$) at least one real or apparent point lattice exists. Points which generate a point lattice in a space group G are called principal points in the sense of Fedorov.

We consider a space group G with point group P. Let $P' < G$ be a maximal point group of G such that P' determines a regular splitting of $O(P, x_0)$ (cf. section 8.2). Following Weissenberg P' is called a principal point group of G (Weis3). Let U be the symmetry support of P'. A point $x_0 \in U$ is called a pricipal point of G in the sense of Weissenberg.

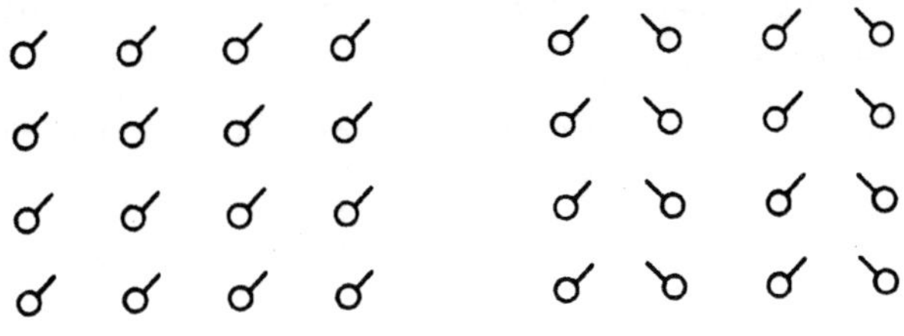

Figure 8.5. a) A real point lattice, b) an apparent point lattice

8.6. Colour groups and colourings

8.6.1. Colour groups

In n-dimensional Euclidean space E^n we consider a set M of objects $\mu_i \subset E^n$ on which a symmetry group H acts transitively,

$$M := \{\mu_1, \mu_2, \ldots\},$$

such that $\mathrm{int}\mu_i \cap \mathrm{int}\mu_j = \emptyset$ for $i \neq j$. The concept of a symmetry group H acting transitively on M can be extended by considering ordered pairs of transformations $(\sigma, P(\sigma))$ acting on M, where $\sigma \in H$, and where $P(\sigma)$ is a transformation, depending on σ only, which acts locally on the objects $\mu_i \in M$. Symbolically we can represent $P(\sigma)$ through a permutation of the colours of the objects. The ordered pairs $(\sigma, P(\sigma))$ form a group by the direct product with composition law

$$(\sigma_1, P(\sigma_1))(\sigma_2, P(\sigma_2)) = (\sigma_1\sigma_2, P(\sigma_1)P(\sigma_2))$$

$$= (\sigma_1\sigma_2, P(\sigma_1\sigma_2)).$$

Following Speiser, we define a colour group (Spei1, appendix):

Definition 8.11: A group of ordered pairs $(\sigma, P(\sigma))$ acting on M is called a colour group if there exists a subgroup $H' < H$ which acts transitively on the objects of the same colour. The action of the ordered pairs $(\sigma, P(\sigma))$ on M is called a perfect colouring of M.

If the number of colours is two then the colour group is also called a black-white group.

As proposed by Koptsik (Kopt1) more general colourings of M can be obtained if we remove the condition that a subgroup $H' < H$ leaves each colour unchanged. Such colourings we call non-perfect and are not considered here.

Suppose the subgroup H' is of index k in H and leaves the colour c_1 unchanged. In general each of the m different colours $c_1, \ldots, c_m$ corresponds to the union of some of the left cosets,

$$H = H' \cup \rho_2 H' \cup \ldots \cup \rho_k H'.$$

The left multiplication of the cosets with a symmetry operation $\sigma \in H$ gives a permutation $P(\sigma)$ of the colours of M,

$$P(\sigma) := \begin{vmatrix} H' & \ldots & \rho_k H' \\ \sigma H' & \ldots & \sigma \rho_k H' \end{vmatrix}.$$

The permutations $P(\sigma)$ give a representation of H by permutations. The group of permutations is a homomorphic image of H. Thus each colour group is determined by an ordered pair (H, H') of a group and a subgroup.

Definition 8.12: Two colour groups (H_1, H_1') and (H_2, H_2') are equivalent and belong to the same colour-group type if there exists an affine transformation $\varphi \in \mathbb{A}(n)$ which maps H_1 onto H_2 and simultaneously maps H_1' onto H_2',

$$H_2 = \varphi H_1 \varphi^{-1}, \quad H_2' = \varphi H_1' \varphi^{-1}.$$

The 212 colour-group types of the three-dimensional crystallographic point groups are given in Table 8.6. They were determined by Harker (Hark1). In Table 8.7 are listed the numbers of colour-group types of the plane groups up to subgroups of index 15. They were determined by Jarrat and Schwarzenberger (Jarr1). A few corrections were added and the list was extended to subgroups up to index 60 by Wieting (Wiet1). The 1191 colour-group types of the three-dimensional space groups with subgroups of index two were determined by Zamorzaev (Zamo1). Harker enumerated the 423 colour-group types with subgroups of index 3 (Hark2).

8.6.2. Perfect colourings

In most works cited above only colour groups were considered but not colourings. Perfect colourings were discussed by Senechal (Sene3). It was already mentioned in section 8.4, that two subgroups of a space group G which are equivalent under $N_A(G)$ may act differently on a set M. Therefore, in order to determine different colourings of M we may have to consider the equivalence of subgroups under H. Moreover the stabilizers of the uncoloured and the coloured set M have to be compared. Let $H_s \geq H$ be the stablizer of the uncoloured set M. We consider an ordered pair (H_s, H') of a group H_s and a subgroup H' of index k. The left coset decomposition of H_s with respect to H' becomes

$$H_s = \rho_1 H' \cup \ldots \cup \rho_k H',$$

where ρ_1 is the identity operation. Under H' the objects $\mu_i = \rho_i \mu_1$, $i=1,\ldots,k$, belong in general to different transitivity classes. Let $St_s(\mu_i) < H_s$ and $St'(\mu_i) < H'$ be the stabilizers of the objects $\mu_i \in M$ in H_s and H' respectively with $St'(\mu_i) \leq St_s(\mu_i)$ of index l_i. If we require that each object $\mu_i \in M$ has only one colour and that H' acts transitively on the objects of the same colour then it follows that l_i of the cosets $\rho_j H'$ must correspond to the same colour hence, the number m of different colours in the perfect colouring of M becomes

$$m = \sum_{i=1}^{k} 1/l_i \qquad (8.6)$$

Using m different colours $c_1, \ldots, c_m$ we perform the colouring of M and determine the stabilizer of the coloured set. The colour-group type assigned to M is the ordered pair (H_s, H'_s). In general we consider two colourings of M, $(H_s, H'_{s,1})$ and $(H_s, H'_{s,2})$ as different if $H'_{s,1}$ and $H'_{s,2}$ are not equivalent under H_s.

Table 8.6. The numbers of colour-group types of the three-dimensional crystallographic point groups

	Index of the subgroup								
Crystal class	2	3	4	6	8	12	16	24	48
C_1-1									
$C_i-\bar{1}$	1								
C_2-2	1								
C_s-m	1								
$C_{2h}-2/m$	3		1						
D_2-222	1		1						
$C_{2v}-mm2$	2		1						
$D_{2h}-mmm$	3		3		1				
C_4-4	1		1						
$S_4-\bar{4}$	1		1						
$C_{4h}-4/m$	3		3		1				
D_4-422	2		2		1				
$C_{4v}-4mm$	2		2		1				
$D_{2d}-\bar{4}2m$	3		3		1				
$D_{4h}-4/mmm$	5		7		5		1		
C_3-3		1							
$C_{3i}-\bar{3}$	1	1		1					
D_3-32	1	1		1					
$C_{3v}-3m$	1	1		1					
$D_{3d}-\bar{3}m$	3	1	1	3		1			
C_6-6	1	1		1					
$C_{3h}-\bar{6}$	1	1		1					
$C_{6h}-6/m$	3	1	1	3		1			
D_6-622	2	1	1	2		1			
$C_{6v}-6mm$	2	1	1	2		1			
$D_{3h}-\bar{6}2m$	3	1	1	3		1			
$D_{6h}-6/mmm$	5	1	5	5	1	5		1	
$T-23$		1	1	1		1			
$T_h-m\bar{3}$	1	1	1	3	1	3		1	
$T_d-\bar{4}3m$	1	1	1	3	1	2		1	
$O-432$	1	1	1	3	1	2		1	
$O_h-m\bar{3}m$	3	1	2	7	3	9	1	5	1
Total	58	17	41	40	17	27	2	9	1

Example 8.5: A black-white tessellation is shown in Figure 8.6. The stabilizer of the tessellation is p4gm, if uncoloured, and c2mm$[\vec{a},\vec{b}]$, if coloured. Two other perfect black-white colourings are possible having stabilizers p4$[2\vec{a},2\vec{b}]$, and pg$[2\vec{a},\vec{b}]$ respectively, both of index 8. Since $l_i=4$, for $i=1,\ldots,8$, it follows from

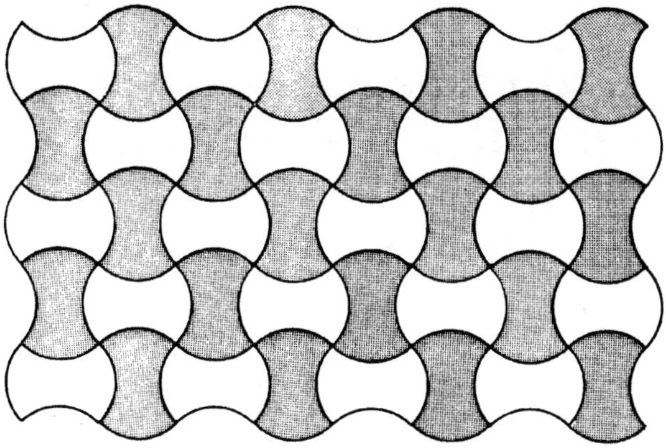

Figure 8.6. Mosaic in the Roman thermae at Nora (Sardinia)

equation 8.6 that the number of colours is m=2. There are other remarkable features in this tessellation. It is tile-, edge-, and vertex-transitive. Moreover, the stabilizer of the set of all vertices is a supergroup of p4gm viz.: $p4mm[\frac{1}{2}(\vec{a}-\vec{b}),\frac{1}{2}(\vec{a}+\vec{b})]$. It is this supergroup which, in this case, determines the equivalence of the colourings.

We note that for the related chessbord 9 perfect black-white colourings are possible. Thereby, two colourings which differ only by a change of black and white are considered to be equivalent. It is a nice exercise to find these 9 perfect colourings.

8.7. Subperiodic groups

We now leave the concept of a discontinuum and consider more general symmetry groups. A symmetry group in n-dimensional space E^n is called a subperiodic group if it contains less than n linearly independent translations. Following Bohm (Bohm1), we denote subperiodic groups by the symbol

$$G^t_{p\cdot\cdot v},$$

Table 8.7. The numbers of colour-group types of the plane groups with subgroups up to index 15

	Index of the subgroup													
Plane group	2	3	4	5	6	7	8	9	10	11	12	13	14	15
p1	1	1	2	1	1	1	2	2	1	1	2	1	1	1
p2	2	1	3	1	2	1	4	2	2	1	3	1	2	1
pm	5	2	10	2	11	2	16	3	12	2	23	2	13	4
pg	2	2	4	2	5	2	7	3	6	2	11	2	7	4
cm	3	2	7	2	7	2	13	3	8	2	17	2	9	4
p2mm	5	1	13	1	9	1	21	2	10	1	25	1	10	2
p2mg	5	2	11	2	11	2	19	3	12	2	26	2	13	4
p2gg	2	1	4	1	4	1	7	2	5	1	9	1	5	2
c2mm	5	1	11	1	8	1	21	2	9	1	22	1	9	2
p4	2	0	5	1	2	0	9	1	4	0	9	1	3	0
p4mm	5	0	13	0	2	0	28	1	3	0	16	0	2	0
p4gm	3	0	7	0	2	0	13	1	3	0	10	0	2	0
p3	0	2	1	0	1	1	0	3	0	0	4	1	0	2
p31m	1	2	1	0	5	0	1	3	0	0	7	0	1	2
p3m1	1	2	1	0	4	0	1	3	0	0	7	0	1	2
p6	1	2	1	0	5	1	1	3	0	0	8	1	1	2
p6mm	3	2	2	0	11	0	3	3	0	0	20	0	1	2
Total	46	23	96	14	90	15	166	40	75	13	219	16	80	34

where t is the dimension of a subspace U which contains t linearly independent translations and p,...,v denote the dimensions of symmetry invariant subspaces $V_p,\ldots,V_v$ which have no translations. For the dimension n it holds that

$$n=t+p+\ldots+v.$$

The fundamental region of a subperiodic group is unbounded. If any of the integers p,...,v is equal or larger than 2 then rotations of infinite order can occur. In the following we only consider subperiodic groups which are subgroups of space groups.

The groups of borders G_1^1 are important in ornamental art. An example is shown in Figure 8.7. We obtain the 7 types of border groups as subgroups of the plane groups. They were determined by Niggli (Nigg3) and are represented in Figure 8.8.

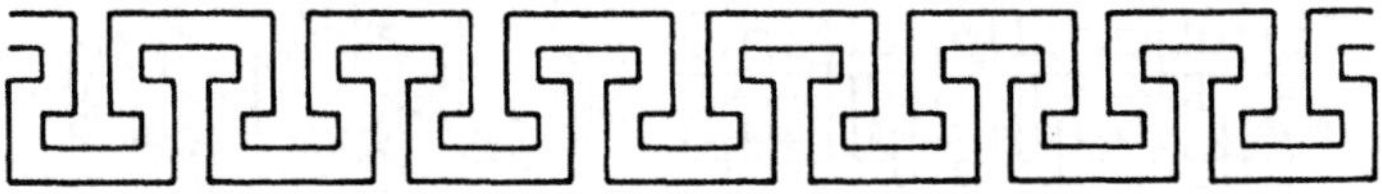

Figure 8.7. Greek border

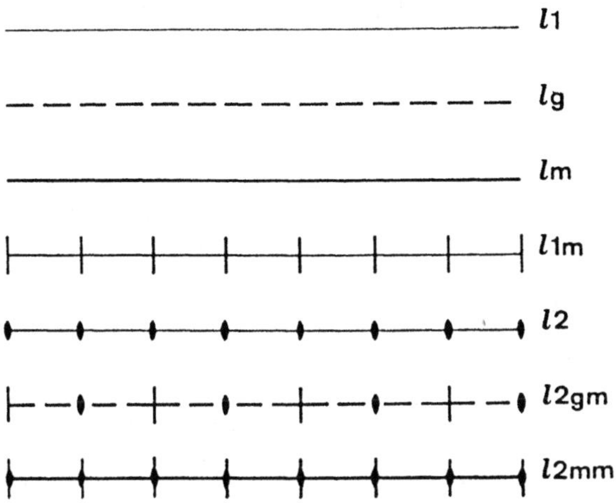

Figure 8.8. Graphical representation of fragments of the symmetry scaffoldings of the 7 border groups

Layer groups G_1^2 have applications in ornamental art and crystallography. Two examples of two-sided plane ornaments are shown in Figure 8.9. The 80 types of layer groups are obtained as subgroups of the three-dimensional space groups. They were independently and almost at the same time determined by C. Hermann (Herm1), Weber (Webe1), and Alexander and K. Herrmann (Alex1).

Other subperiodic groups have been determined and extended to several colours. The numbers of types of subperiodic groups given in Table 8.8 are taken form Zamorzaev and Palistrant (Zamo3). We note that their numbers of three- and four-coloured subperiodic groups are incorrect.

Figure 8.9. Two-sided plane ornaments

Table 8.8. The numbers of types of subperiodic groups according to Zamorzaev and Plistrant

Subperiodic group		1-coloured	2-coloured
crystal classes in E^2	G_2^0	10	11
crystal classes in E^3	G_3^0	32	58
linear groups	G_0^1	2	3
one-sided border groups	G_1^1	7	17
two-sided border groups	$G_{1,1}^1$	31	117
rod groups	G_2^1	75	224
plane groups	G_0^2	17	46
layer groups	G_0^2	80	368
space groups	G_0^3	219	1191

8.8. Historical remarks

Space groups with proper motions only – those which conserve the chirality of an object – were investigated first by the mathematician Camille Jordan (1838-1922) (Jord1). Using the results of Jordan, Sohncke later determined in a systematic way the 54+11 three-dimensional space-group types with proper motions only. (The notation 54+11 indicates that there are 11 enatiomorphic pairs which are mirror images of each other). Originally Sohncke listed one space-group type twice. Soon however, it became evident that also improper motions are important such as mirror planes, glide planes, and rotoinversions, which change the chirality of a molecule within a crystal. This was emphasized in 1885 by Pierre Curie (Curi1) and in 1887 by Leonhard Wulff (Wulf1). This latter work led Sohncke to extend his theory on crystal structure (Sohn4) but he did not complete the enumeration of all space-group types. In the meantime Fedorov, using the results of Sohncke, derived 218+11 space-group types. Independently the mathematician Schönflies, prompted by Felix Klein, obtained 216+11 space-group types. Only after comparing their results the correct number of 219+11 space-group types was established (Schö3). In subsequent years the space groups were again derived by William Barlow (1845-1934) (Barl1). The interesting history of the discovery of the space groups was described by Burckhardt (Burc3).

The regular point systems in the plane were investigated by Sohncke in 1874 in a purely geometric way without using group theory. He found 13 different construction types. These correspond exactly to the 13 types of stabilizers of regular point systems in the plane. By specializing the construction conditions Sohncke also found all possible orbit types in the plane. These results are summerized in Table 8.5. In 1891 Fedorov derived the 17 plane-group types from the three-dimensional space groups (Fedo4). This result remained unknown and in 1924 the plane-group types were again determined by Georg Polyá (Poly1).

In 1900 the mathematician David Hilbert (1862-1943) presented ten of his 23 unsolved problems at the international mathematical congress in Paris. In the first part of the 18th problem Hilbert asked whether, in n-dimensional Euclidean space, only a finite number of non-isomorphic space groups with finite fundamental region exists (Hilb1). In 1910 Ludwig Bieberbach proved that for any dimension this number is finite (Bieb1) and in 1911 Ferdinand Georg Frobenius (1849-1917) showed that the number of space groups which are not affine-equivalent is also finite (Frob1). Finally Bieberbach proved that in n-dimensional space non-affine space groups are also non-isomorphic (Bieb2).

The methods used by Frobenius and Bieberbach were futher developed by Johann Jakob Burckhardt in a series of papers, and in his influencel book (Burc1, Burc2). An analogous algorithm was also developed by Hans Zassenhaus (Zass1). The application of group extensions to the determination of space groups was developed by Edgar Ascher and Alois Janner (Asch1, Asch2).

The investigation of four-dimensional space groups was started in 1930 by Heinrich Heesch (Hees1) while investigating three-dimensional black-white groups. In 1978 a complete list of the 4789+112 four-dimensional space groups was achieved by Harold Brown, Rolf Bülow, Joachim Neubüser, Hans Wondratschek, and Hans Zassenhaus (Brow1).

Crystallographic orbits were investigated by Paul Niggli who synonymously used the term "lattice complex" (Nigg1,Nigg2). A systematic investigation of crystallographic orbits was done by Hans Wondratschek (Wond2). The concept of lattice complex was developed in a different way by several authors (cf. Burz1).

Coloured plane tessellations can be found already in ancient civilisations. Particularly the ancient Agyptian and Islamic culture have a very rich ornamental art (the reader is referred to Jone1). In our century the ornamental art was revived again by the Dutsch artist Mauritius Cornelius Escher (1898-1972) whom we owe many beautiful tessellations (Erns1,Fuij1). The term "black-white" groups was introduced by L. Weber in 1929 (Webe1). The generalization to more than two colours and the theory of colour groups was published in 1956 by the mathematician Andreas Speiser (1885-1970). Uncorrectly this development is often attributed to van der Waerden and Burckhardt (Waer2). Indepentently polycolour symmetry was also investigated by N. V. Belov and E. N. Belova (Belo1).

Examples for subperiodic groups are frequent in ancient ornamental art. The border groups were derived in 1926 by Paul Niggli.

9. Space Partitions

The investigation of space partitions permits us to get a better insight into the discontinuum. The most important result of this chapter will be the regularity condition which is discussed in section 9.4.

9.1. Definitions

9.1.1. Space partition

In Euclidean space E^n we consider a family of n-dimensional tiles

$$\{T_\nu \mid \nu \in I\},$$

where each tile T_ν is a bounded closed subset of Euclidean space, $T_\nu \subset E^n$, and I is an infinite set of indices.

Definition 9.1: A family of tiles $\{T_\nu | \nu \in I\}$, is called a space partition

1. if the union of all tiles fills space completely,

$$\bigcup_{\nu \in I} T_\nu = E^n.$$

2. if the intersection of the interiors of any two tiles is empty

$$\text{int}T_i \cap \text{int}T_j = \emptyset, \quad i \neq j.$$

Following Grünbaum and Shephard (Grün2), the space partition is called monohedral if all its tiles are directly or mirror congruent to a prototile T_0. It is called isohedral if there exists a space group G which acts transitively on the tiles. Synonymous to isohedral also regular or homogeneous is used. An isohedral space partition necessarily is monohedral. A space partition is called face-to-face if the intersection of any two tiles is either empty or a d-face of each. It is called normal if there exist two balls, B_1 of radius $\rho_1>0$, and B_2 of finite radius ρ_2 such that for each tile T_v it holds that $B_1 \subset T_v \subset B_2$. From the definition of a space group (cf. section 1.4) it follows that an isohedral space partition is normal. A space partition is called locally finite if around every point in E^n there exists a ball of radius $\rho>0$ which is part only of a finite number of tiles. Clearly normal includes locally finite.

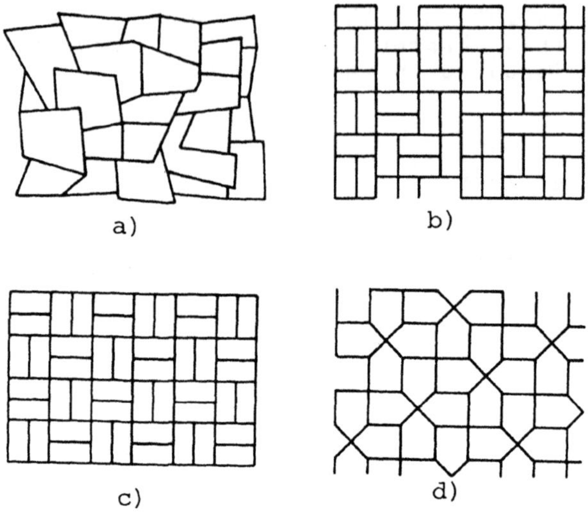

Figure 9.1. Normal space partitions: a) general, b) monohedral, c) isohedral, d) isohedral and face-to-face

9.1.2. Convex polytopes

In what follows we assume that each tile is a n-dimensional convex polytope P called, following Fedorov, a stereotope (Fedo2). It is obtained through the intersection of a finite number N of open half-spaces H_i^+ (cf. section 2.1),

$$P^+ = \bigcap_{i=1}^{N} H_i^+ .$$

We denote by P the closure of P^+, $P = \mathrm{cl}P^+$.

Convex means that for any two points $p_1, p_2 \in P$ also every point of the closed line segment $[p_1, p_2]$ belongs to P. Each point $x \in [p_1, p_2]$ can be represented as

$$\vec{x} = \lambda\vec{p}_1 + (1-\lambda)\vec{p}_2, \quad 0\leq\lambda\leq 1.$$

This equation is equivalent to

$$\vec{x} = \lambda_1\vec{p}_1 + \lambda_2\vec{p}_2, \quad 0\leq\lambda_1,\lambda_2; \ \lambda_1+\lambda_2=1.$$

The vertices v_i, $i=1,\ldots,N_0$, of the polytope P are the extreme points of P (cf. section 2.2). It follows that each point $x\in P$ can be represented as

$$\vec{x} = \sum_{i=1}^{N_0} \lambda_i\vec{v}_i, \quad 0\leq\lambda_i; \ \sum_{i=1}^{N_0} \lambda_i=1.$$

9.1.3. Characterization of polytopes

In the topological morphology of polytopes each polytope is characterized by its hierarchical structure (cf. section 2.4). Metrical properties such as the d-dimensional volume of a d-face, or the dihedral angles between the facets are not considered.

Definition 9.2: Two polytopes P_1 and P_2 in n-dimensional space are called isomorphic and belong to the same combinatorial type if P_1 and P_2 have isomorphic hierarchical structures.

The isomorphism of two polytopes is not easily recognized. In what follows a sequence of gradually finer classifications is given.

A rough classification of a polytope is obtained by the numbers N_d of d-faces, $d=0,\ldots,n$.

A finer classification is obtained by the d-subordination symbol, $d=1,\ldots,n$. Let $M:=\{F_1,\ldots,F_{N_d}\}$ be the set of d-faces of a polytope P. The d-subordination symbol

$$f_{1\,m_1}\, f_{2\,m_2} \ldots f_{k\,m_k}$$

gives the numbers m_i of those d-faces $F_j\in M$ which have subordinated f_i (d-1)-faces, $i=1,\ldots,k$. It follows that the number N_d of d-faces is given by

$$N_d = \sum_{i=1}^{k} m_i.$$

In a similar way we define the d-superordination symbol, $d=0,\dots,n-1$,

$$f_1^{m_1} f_2^{m_2} \dots f_l^{m_l},$$

where m_i is the number of those d-faces $F_j \in M$ which have superordinated f_i (d+1)-faces, $i=1,\dots,l$.

The numbers N_d of d-faces of a polytope are connected by Euler's relation (Eule1). In three-dimensional space this relation is given by

$$N_0 - N_1 + N_2 = 2.$$

In n-dimensional space the relation was given by Schläfli (for a proof see Schl1, Bron1),

$$\sum_{i=0}^{n-1} (-1)^i N_i = 1 - (-1)^n.$$

A unique classification of a polytope P is given by the unified polytope scheme. It is obtained by labeling the d-faces of a polytope in a unified way. Thereby, we use the fact that there exists a partial order among the elements of a polytope and that each (n-2)-face is exactly contained in two facets (cf. section 2.4). For $d>1$ the (d-1)-faces $F'_1,\dots,F'_k,F'_{k+1}=F'_1$ contained in a d-face F can be cyclically ordered by the condition that for $i=1,\dots,k$, $F'_i \cap F'_{i+1}=F''_i$, where F''_i is a (d-2)-face of F. We say a d-face to be ordered or unordered if its subordinated (d-1)-faces are cyclically ordered or not. An ordered pair ($F' \subset F$) of a (d-1)-face F' contained in a d-face F is called a <u>d-flag.</u> The unified labeling of a polytope P is proposed with the following algorithm:

1. We choose among the facets of P which contain the largest numbers of (n-2)-faces a (n-1)-flag, say ($F'_1 \subset F_1$), where the labeling shall start, together with the sense of the cyclic enumeration. The (n-2)-faces contained in F_1 are labeled in cyclic order and the cycle for F_1 is written down.

2. The enumeration continues in that unordered facet which contains the (n-2)-face with the smallest number. Starting from this (n-2)-face, but in the oposite sense, the cyclic labeling is performed and the cycle is written down. Thereby, those (n-2)-faces which were not labeled priviously are labeled by successive

numbers. We repeat step 2 until the cycles for all facets are written down. Finally the facets are labeled in the order of their occurence during the enumeration process.

3. Suppose the d-faces $F_1,\ldots,F_{N_d}$ have been labeled. The enumeration of the (d-1)-faces F' is performed for each F_i, $i=1,\ldots,N_d$. We choose as starting d-flag $(F'_s \subset F_i)$ such that for $F'_s=F_i \cap F_j$, j is minimal. Given F'_s, the sense of the cyclic enumeration is determined by the condition that for $F'_{s+1}=F_i \cap F_k$, k is minimal. Now the (d-1)-faces of F_i are labeled in cyclic order. Thereby, those (d-1)-faces which were not labeled previously are labeled by successive numbers. The cycle for F_i is written down.

4. Step 3 is repeated until the edges (1-faces) have been labeled. This completes the polytope scheme. We recall that each edge contains exactly two vertices. Since the sense of enumeration in each cycle is given, every edge has a relative orientation with respect to that cycle. Thus relative to a cycle one vertex of every edge can be uniquely assigned as the first vertex. The first vertices of each edge are now labeled in the same cyclic order as the edges. Thereby, a vertex which was not labeled priviously is labeled by a successive number.

5. The polytope scheme thus obtained depends only upon the starting (n-1)-flag and the sense of cyclic enumeration. In order to make the polytope scheme unique we write down the schemes for each (n-1)-flag and both directions of cyclic enumeration, but only those schemes are retained which have the longest second cycle. Amongst these all those with the longest third cycle and further with the longest i^{th} cycle are retained. Amongst the remaining schemes one is chosen wherein a smaller number first occurs.

With this algorithm a unique polytope scheme can be assigned to each convex polytope. Two isomorphic types yield always the same scheme. If the sense of enumeration is defined, enantiomorphic polytopes are distinguished. If both directions of enumeration are considered, the number of equal polytope schemes corresponds to the order of the automorphism group aut(P) of the hierarchical structure of P.

The unified polytope scheme allows us to order the elements of P. We say F'<F if in the polytope scheme the cycle of F' is situated after the cycle of F.

For two-dimensional polygons the polytope scheme is trivial. The combinatorial type is uniquely characterized by the number of edges of a polygon.

The polytope schemes for three-dimensional polyhedra are easily set up. In spaces of higher dimensions the polytopes have a large number of elements therefore, the application of this algorithm is limited, even for large computers. Compared to the two- and three-dimensional space little is known on polytopes in higher dimensions. In Table 9.1 are listed the numbers of combinatorial types of three-dimensional polyhedra with up to 11 faces (Enge3).

Table 9.1. Numbers of combinatorial types of polyhedra in three-dimensional space with N_2 faces and N_0 vertices

N_0 \ N_2	4	5	6	7	8	9	10	11
4	1							
5		1	1					
6		1	2	2	2			
7			2	8	11	8	5	
8			2	11	42	74	76	38
9				8	74	296	633	768
10				5	76	633	2635	6134
11					38	768	6134	25626
12					14	558	8822	64439
13						219	7916	104213
14						50	4442	112082
15							1404	79773
16							233	36528
17								9714
18								1249
Total	1	2	7	34	257	2606	32300	440564

9.1.4. Characterization of space partitions

The combinatorial type of stereotopes, in general, does not determine the type of space partition. There exist stereotopes congruent copies of which can fill space in infinite many different ways. Even if we require the space partition to be isohedral, then different types of space partitions are possible. In the plane, using a brick with edge ratio 1:2 as a prototile, 9 different isohedral tessellations can be obtained which are shown in Figure 9.2, a-i. Thereby, tessellations which are obtained by interchanging the short with the long edge of the brick are not considered as different.

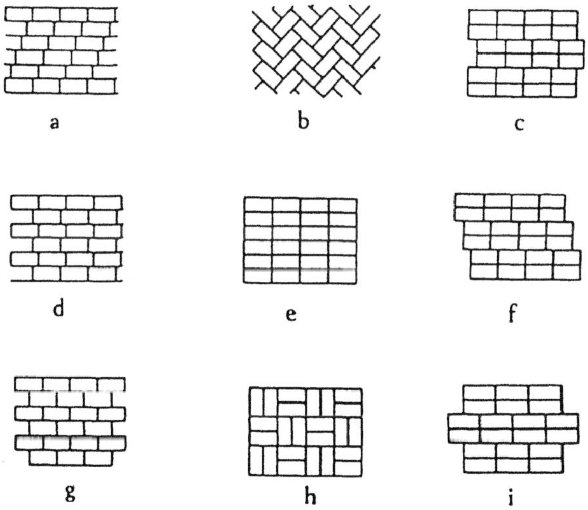

Figure 9.2. Different isohedral plane partitions with bricks

The topological classification of a general space partition is cumbersome and is beyond our scope. In what follows we restrict ourselves to isohedral face-to-face space partitions.

Let the stereotope P_0 be a prototile of an isohedral face-to-face space partition. In order to obtain a topological characterization of such a space partition we propose an adjacency scheme with the following algorithm:

1. The unified polytope scheme allows us to order the faces $F_0 \in P_0$, $F_{01}, \ldots, F_{0\nu}$. We assign to each face $F_{0i} \in P_0$, $i=1,\ldots,\nu$, all faces $F_j \in P_j$, where P_j intersects P_0 in F_{0i}, $P_0 \cap P_j = F_{0i}$. By $N(F_{0i})$ we denote the number of such stereotopes.

2. To make this adjacency scheme unique we have also to consider the automorphism groups $\mathrm{aut}(P_0)$ and $\mathrm{aut}(P_j)$. We require that $N(F_{01})$ is maximal. Among the remaining schemes we retain those with maximal $N(F_{02})$ etc. Second we require that the labels of all faces F_j assigned to F_{0i} are minimal for $i=1,\ldots,\nu$.

This algorithm allows us to classify isohedral face-to-face space partitions according to their topological type with the following definition.

<u>Definition 9.3:</u> Two isohedral face-to-face space partitions τ and τ' are isomorphic and belong to the same topological type of isohedral face-to-face space partitions if there exists a bijective mapping φ: $\tau \to \tau'$ such that for $T_i \in \tau$, it holds that T_i and $\varphi(T_i)$ have identical polytope shemes and adjacency shemes.

We note that the adjacency scheme does not determine the isohedral type, that is, the type of stabilizer G_s of the space partition. In general, under a continuous deformation of the space partition also the stablilizer G_s changes.

A finer classification is obtained if we consider the stabilizer G_s of the space partition and the way G_s acts on the tiles.

Definition 9.4: Two isohedral face-to-face space partitions τ and τ' are isomorphic and belong to the same isohedral type of face-to-face space partition

1. if τ and τ' belong to the same topological type of isohedral face-to-face space partition.

2. if τ and τ' have equivalent stabilizers G_s and G_s' respectively, that is, if there exists an affine mapping φ such that $G_s' = \varphi G_s \varphi^{-1}$.

3. If the corresponding d-faces $F \in \tau$ and $F' \in \tau'$, $d=0,\ldots,n$, have equivalent stabilizers St(F), St(F') in G_s and G_s' respectively such that $St(F') = \varphi St(F) \varphi^{-1}$.

9.2. Dirichlet domain partitions

The construction of the Dirichlet domains of a regular point system in Euclidean space E^n is a very efficient method to obtain isohedral face-to-face space partitions which are of particular interest in crystallography.

9.2.1. Dirichlet domain partitions in the plane E^2

The isohedral face-to-face tessellations in the plane E^2 are well known. An important property is reveald by the following theorem due to Reinhardt (Rein1).

Theorem 9.1: In a normal face-to-face tessellation in E^2 with stereogons, all of which are of the same combinatorial type, only three-, four-, five-, or six-gons are possible.

Proof: Let p be the number of edges of a stereogon. Clearly $p \geq 3$. We determine the least upper bound for p. By the normal condition, there exist two fixed radii, ρ_1, ρ_2. For any circle S^1 in E^2 of radius $\nu\rho_2$ we denote by a_ν the number of all stereogons which intersect S^1 and by A_ν all stereogons which lie in the interior of S^1 but do not touch S^1. It follows that

$$\frac{a_\nu}{A_\nu} < \frac{\rho_2 \nu}{4\rho_1(\nu-1)^2} \to 0, \quad \text{for } \nu \to \infty.$$

Hence, the topological properties of the tessellation are determined by the stereogons in the interior of S^1 only, as ν goes to infinity. From this, and because the tessellation is face-to-face, it follows that the corresponding edge graph fulfils Euler's relation,

$$N_0 - N_1 + N_2 = 2.$$

Every edge belongs exactly to two stereogons, each having p vertices hence,

$$2N_1 = N_2 p.$$

The number of vertices in an edge graph is maximal if in every vertex exactly three edges meet. Since every edge has two vertices it follows that in this case

$$N_0 = \tfrac{2}{3}N_1 = \tfrac{1}{3}N_2 p.$$

Substitution of N_0 and N_1 in Euler's relation gives

$$\tfrac{1}{3}N_2 p - \tfrac{1}{2}N_2 p + N_2 = 2.$$

From this follows that $p=6(1-2/N_2)$. For $N_2 \to \infty$ the result $p=6$ follows which proves the theorem.

The normal condition in theorem 9.1 is essential because each face-to-face tessellation in the plane with p-gons, where $p>6$, necessarily is not normal as was proved by Laves (Lave2).

Remarkably theorem 9.1 does not require an isohedral tessellation. However, for the derivation of the topological types of tessellations we replace the normal condition by the stronger isohedral condition in order to keep their number finite. The adjacency scheme for a plane tessellation τ is trivial. Let P_0 be a stereogon of τ. It is sufficient to specify for each vertex $v_i \in P_0$ the number $N(v_i)$ of stereogons P_j for which $P_j \cap P_0 = v_i$. We note that the number of stereogons meeting in v_i is $N(v_i)+3$. The adjacency scheme has the form

$$N(v_1)-N(v_2)-\ldots-N(v_p).$$

In order to get a complete description of the type of tessellation the adjacency scheme is extended by the polytope scheme (cf. section 9.1.3). Thus we obtain

$$p|N(v_1)-N(v_2)-\ldots-N(v_p).$$

We call a stereotope in E^n maximal, if it cannot be derived from another stereotope in E^n by reduction processes. A primitive stereotope in the sense of Voronoi (cf. Definition 3.4) necessarily is maximal.

The topological types of isohedral face-to-face tessellations in the plane E^2 can be determined using this concept of maximal stereogons as follows:

1. In the plane E^2 there exists one maximal topological type of isohedral face-to-face tessellation, viz.: 6|0-0-0-0-0-0 (cf. Figure 9.3,l). This type allows several different isohedral face-to-face tessellations. Actually, every centrosymmetric hexagon is a primitive parallelogon. In space partitions of higer dimensional spaces the maximal stereotopes, in general, are not primitive parallelotopes and they not necessarily allow an isohedral tiling.

2. All the other topological types of isohedral face-to-face tessellations having stereogons with less edges can be obtained by edge reduction, that is, an edge E_{ij} is shrunk up until both subordinated vertices v_i and v_j fall together. Thereby, both stereogons which are adjacent to E_{ij} are reduced by one edge. By the valence $V(v_i)$, we denote the number of edges meeting in v_i. The valence of the new vertex becomes $V(v_i)+V(v_j)-2$. In order to obtain the reduced types we can make either single edge reductions such that each p-gon is reduced to a (p-1)-gon, or else we can make simultaneous edge reductions such that a complex of one or several stereogons, at a time, shrinks up to a vertex. Thereby, all p-gons P_j which share an edge with this complex are reduced to (p-1)-gons. The isohedral condition requires that each stereogon $P_i \in \mathcal{T}$ has the same adjacency scheme. This poses severe restrictions on the possible edge reductions.

Example 9.1: We derive the 11 topological types of isohedral face-to-face tessellations in the plane E^2. From the maximal type 6|0-0-0-0-0-0 we obtain by single edge reduction the types 5|1-0-1-0-0 and 5|1-1-0-0-0 shown in Figures 9.3,h,i. By reduction of hexagons we get the

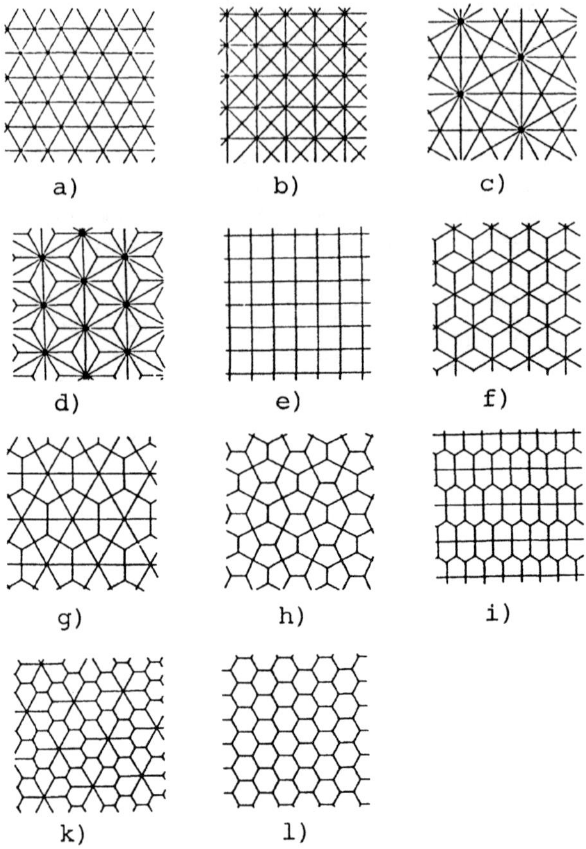

Figure 9.3. The 11 topological types of isohedral face-to-face tessellations in the plane E^2

types 5|3-1-0-0-0 (Figure 9.3,k), 4|3-0-3-0 (Figure 9.3,f), and 3|3-3-3 (Figure 9.3,a). From the types 5|1-1-0-0-0 and 5|1-0-1-0-0 we derive by single edge reduction the type 4|1-1-1-1 (Figure 9.3,e). We note that the reduction of pentagons does not result in a monohedral tessellation. From the type 5|3-1-0-0-0 we obtain by single edge reduction the type 4|3-1-0-0

(Figure 9.3,g). From the type 4|1-1-1-1 we derive either by single edge reduction, or by reduction of quadrangles the type 3|5-5-1 (Figure 9.3,b). From the type 4|3-1-0-0 we obtain the type 3|4-3-1 (Figure 9.3,c) by simultaneous reduction of complexes of six stereogons, at a time, which form together a hexagon. Finally from the type 4|3-0-3-0 we get the type 3|9-9-1 (Figure 9.3,d) by simultaneous reduction of complexes of three stereogons, at a time, which form together a hexagon.

The result of these reductions may be summerized in the following theorem due to Laves (Lave2).

Theorem 9.2: In the plane E^2 there exist 11 topological types of isohedral face-to-face tessellations.

We note that each of these 11 topological types of isohedral face-to-face tessellations can be realized by a Dirichlet domain partition (Lave1).

Considering the stabilizer of the tessellation and the symmetry of the tiles a refined classification into isohedral types of face-to-face tessellations was performed by Grünbaum and Shephard (Grün1).

Theorem 9.3: In the plane E^2 there exist 47 isohedral types of face-to-face tessellations.

It is still unknown, if each of these 47 isohedral types of face-to-face tessellations can be realized by a Dirichlet domain partition.

Also more general monohedral tessellations have been considered. We note that every triangle and every convex quadrangle can tile the plane monohedrally. Also every centrosymmetric hexagon has this property. Moreover, there exist three families of convex hexagons, discovered by Reinhardt (Rein1), which show this property. Monohedral tessellations with pentagons are more intriguing. Up till now 13 families of convex pentagons were discovered, but there is no prove tat this list is complete (for details the reader is referred to Grün2, Grün3).

9.2.2. Dirichlet domain partitions in E^3

Only a few general results on space partitions in E^3 are known. The maximal number of faces of a stereohedron P, which allows a normal face-to-face space partition with tiles, all of which are combinatorially equivalent to P, is still unkown.

Delone and Sandakova, considering isohedral space partitions in n-dimensional space E^n, proved the following theorem which we state without proof (for a proof see Delo3, Štog1).

<u>Theorem 9.4:</u> Let τ be an isohedral space partition in E^n with stabilizer G_s, having a stereotope P as prototile. The number of facets of P is at most

$$p \le 2^n(1+h)-2,$$

where h is the order of the crystallographic point group belonging to the stabilizer G_s.

In three-dimensional space h=48 is maximal hence, $p\le 390$. To our knowledge this upper bound is much too high.

In three-dimensional space E^3 the primitive parallelohedron is a maximal stereohedron (cf. sections 3.5, 9.3). But more complicated maximal stereohedra exist in E^3 which are not yet known. Most results available up till now are obtained by calculating the Dirichlet domains of crystallographic orbits (Löck1, Koch1, Enge1, Enge2). The calculations required are performed as follows:

Let G be a space group with origin O and lattice basis $\vec{a}_1$, $\vec{a}_2$, $\vec{a}_3$. First we determine the radius R' of a maximal interstitial ball of the lattice T^3, generated by $\vec{a}_1$, $\vec{a}_2$, and $\vec{a}_3$. Since every crystallographic orbit $O(G,x_0)$ is the union of a finite set of lattices, each parallelly placed in space, it follows that a maximal interstitial ball of $O(G,x_0)$ has a radius $R\le R'$. For some point x_0 we determine the crystallographic orbit $O(G,x_0)$ within a ball of radius 2R' and center x_0. Since $R\le R'$, it follows from theorem 2.2, that these points determine the Dirichlet domain $D(x_0)$ (cf. sections 2.2, 2.4). If the coordinates of the reference point x_0 are continuously changed, then the elements of the Dirichlet domain $D(x_0)$ undergo a continuous alteration. The region of the parameter space that corresponds to the same combinatorial type of Dirichlet domains is called the existence region for $D(x_0)$. Such existence regions can be zero, one-, two-, or three-dimensional. Two contiguous d-dimensional existence regions are separated by a (d-1)-dimensional existence region, called a boundary region. The corresponding Dirichlet domain is called a limiting type.

A two-dimensional boundary region is characterized by the condition that at least four faces of $D(x_0)$ intersect in a common vertex $v_h \in D(x_0)$. This provides an approach to calculate the boundary region. Let H_i° be the plane which bisects the face-vector $\vec{f}_i := \overrightarrow{x_0 x_i}$, where $x_i = (S_i, \vec{s}_i) x_0$, $(S_i, \vec{s}_i) \in G$. The condition that H_i contains a vertex v_h is given by

$$\vec{f}_i \bullet \vec{v}_h = \vec{f}_i \bullet \vec{q}_i ,$$

where q_i is any point of H_i°. The condition that four different planes $H_1^\circ, \ldots, H_4^\circ$, intersect in the vertex v_h requires that the corresponding equations are linearly dependent. This means that the determinant Δ has to be zero,

$$\Delta := \begin{vmatrix} \vec{f}_1 & -\vec{f}_1 \bullet \vec{q}_1 \\ \vec{f}_2 & -\vec{f}_2 \bullet \vec{q}_2 \\ \vec{f}_3 & -\vec{f}_3 \bullet \vec{q}_3 \\ \vec{f}_4 & -\vec{f}_4 \bullet \vec{q}_4 \end{vmatrix} = 0 .$$

This determinant provides the equation of a two-dimensional boundary region. We express this equation through the coordinates ξ, η, ζ of the reference point x_0.

Theorem 9.5: In three-dimensional space E^3 every two-dimensional boundary region is determined by a polynom of highest degree four in ξ, η, and ζ.

Proof: We choose $\vec{q}_i := \frac{1}{2}(\vec{x}_0 + \vec{x}_i)$. As $\vec{f} = \vec{x}_i - \vec{x}_0$ it follows that

$$\vec{f}_i \bullet \vec{q}_i = \tfrac{1}{2}(|\vec{x}_i|^2 - |\vec{x}_0|^2).$$

Since the symmetry operation $(S_i, \vec{s}_i) \in G$ is an isometry it holds that $|S_i \vec{x}_0|^2 = |\vec{x}_0|^2$ hence,

$$\vec{f}_i \bullet \vec{q}_i = (S_i \vec{x}_0) \bullet \vec{s}_i + \tfrac{1}{2}|\vec{s}_i|^2 = \delta_i .$$

Thus δ_i is linear in ξ, η, and ζ which proves the above theorem.

Example 9.2: We determine the isohedral types of Dirichlet domain partitions for the space group G:=I23 (the standard setting of the International Table, Vol. A, is used). It is sufficient to consider as parameter space the fundamental region of the corresponding Euclidean normalizer. From Table 8.3 we take $N_{IE}(I23)=Im\bar{3}m$. The fundamental region is the tetrahedron determined by the four vertices (0,0,0), ($\frac{1}{2}$,0,0), ($\frac{1}{2}$,$\frac{1}{2}$,0), and ($\frac{1}{4}$,$\frac{1}{4}$,$\frac{1}{4}$). We choose a grid, having a sufficiently small grid interval, and determine the Dirichlet domain at each grid point. Knowing the Dirichlet domains on both sides of a two-dimensional boundary region, we can deduce the limiting Dirichlet domain and calculate the equation of the boundary region. The results of these calculations are given in Table 9.2. There the type of space partition is denoted by a short symbol, viz.: f.v-m, where f, v are the numbers of faces and vertices respectively of the Dirichlet domain, and m is an integer specifying its combinatorial type.

Table 9.2. The isohedral types of Dirichlet domain partitions in the space group I23

Type	2-subordination	dim	Wyckoff position	existence region
4.4-1	3_4	0	12e	0.5,0.25,0
6.5-1	3_6	1	24f	$0.5-\xi,0.25+\xi,0$ $(0<\xi\leq 0.125)$
6.8-2	4_6	0	8c	0.25,0.25,0.25
8.6-1	3_8	1	24f	$0.25,0.25,\zeta$ $(0\leq\zeta<0.25)$
9.8-2	$3_6 4_3$	1	24f	$\alpha)$ $(\xi=\eta)$
9.8-3	$3_6 4_3$	1	24f	$0.5-\xi,0.25,\xi$ $(0<\xi<0.25)$
9.10-2	$3_2 4_7$	1	24f	$0.375-\xi,0.375-\xi,2\xi$ $(0<\xi<0.125)$
9.12-1	$4_8 6_1$	2	24f	ξ,η,ζ
10.7-1	3_{10}	1	24f	$\beta)=\delta)$ $(\zeta=0)$
	3_{10}	1	24f	$\gamma)$
10.9-1	$3_6 4_4$	1	24f	$\xi,\xi,0$ $(0.25<\xi<0.375)$

Table 9.2. (continued)

Type	2-subordination	dim	Wyckoff position	existence region
10.10-4	$3_4 4_6$	1	24f	$\xi, 0.25, 0$ $(0.25<\xi<0.5)$
10.11-1	$3_4 4_4 5_2$	2	24f	$\xi, \eta, 0$
10.11-2	$3_4 4_4 5_2$	1	24f	$\xi, \xi, 0$ $(0.375<\xi<0.5)$
	$3_4 4_4 5_2$	2	24f	$\xi, \eta, 0$
10.12-1	4_{10}	2	24f	ξ, ξ, ζ
11.13-2	$3_4 4_4 5_2 6_1$	2	24f	$\xi, 0.75-\xi-\zeta, \zeta$
11.13-3	$3_5 4_1 5_5$	2	24f	$\xi, \eta, 0$
11.13-4	$3_1 4_9 5_1$	2	24f	$\xi, \eta, 0$
11.14-1	$3_2 4_7 6_2$	2	24f	ξ, ξ, ζ
11.14-2	$3_2 4_5 5_4$	2	24f	ξ, ξ, ζ
12.10-1	$3_8 4_4$	0	6b	$0.5, 0, 0$
12.13-1	$3_4 4_6 5_2$	2	24f	$\xi, 0.25, \zeta$
13.14-1	$3_8 5_4 6_1$	1	12e	$0.5, \eta, 0$ $(0<\eta<0.25)$
13.15-1	$3_6 4_3 5_2 6_2$	2	24f	δ)
13.15-2	$3_4 4_6 5_2 6_1$	2	24f	δ)
13.15-3	$3_4 4_6 5_2 6_1$	2	24f	α)
13.17-1	$4_{12} 8_1$	1	12d	$\xi, 0, 0$ $(0<\xi<0.5)$
13.18-1	$3_2 4_7 6_4$	3	24f	
14.17-5	$3_7 4_3 5_1 6_1 7_2$	2	24f	β)
14.17-6	$3_4 4_6 5_2 6_2$	2	24f	β)
14.20-7	$3_2 4_7 6_5$	3	24f	
14.20-8	$4_9 5_2 6_3$	3	24f	
14.24-1	$4_6 6_8$	0	2a	$0, 0, 0$
15.17-1	$3_8 4_4 6_2 8_1$	2	24f	ε)
15.20-1	$3_4 4_7 6_3 8_1$	3	24f	
15.22-1	$3_6 4_3 5_2 6_1 7_3$	3	24f	
15.26-1	$4_6 6_9$	1	8c	ξ, ξ, ξ $(0<\xi<0.25)$
16.24-7	$3_6 4_2 5_2 6_4 8_2$	3	24f	
17.24-1	$3_8 4_3 5_2 6_1 8_2 10_1$	3	24f	

α) $3-10\xi+8\xi^2+8\xi\eta-4\eta+8\eta\zeta-2\zeta+8\zeta^2=0$

β) $8\xi^3\zeta-2\xi^2\eta-8\xi^2\zeta+\xi\eta+4\xi\eta^2-8\xi\eta^3+8\xi\eta\zeta+16\xi\eta\zeta^2+2\xi\zeta-2\xi\zeta^2$
$-2\eta^2+4\eta^3-8\eta^3\zeta+6\eta^2\zeta-8\eta^2\zeta^2-2\eta\zeta-4\eta\zeta^2+\zeta^2=0$

γ) $\delta\cap\beta$

δ) $\xi-4\xi\zeta-2\eta+4\eta^2+\zeta=0$

ε) $8\xi^2\eta-16\xi^2\eta^2+32\xi^2\eta\zeta-12\xi^2\zeta-10\xi\eta+28\xi\eta^2-16\xi\eta^3-36\xi\eta\zeta+14\xi\zeta-8\xi\zeta^2$
$+16\xi\zeta^3+3\eta-10\eta^2+8\eta^3-16\eta^3\zeta+16\eta^2\zeta+4\eta\zeta+4\eta\zeta^2-3\zeta+2\zeta^2-8\zeta^3=0$

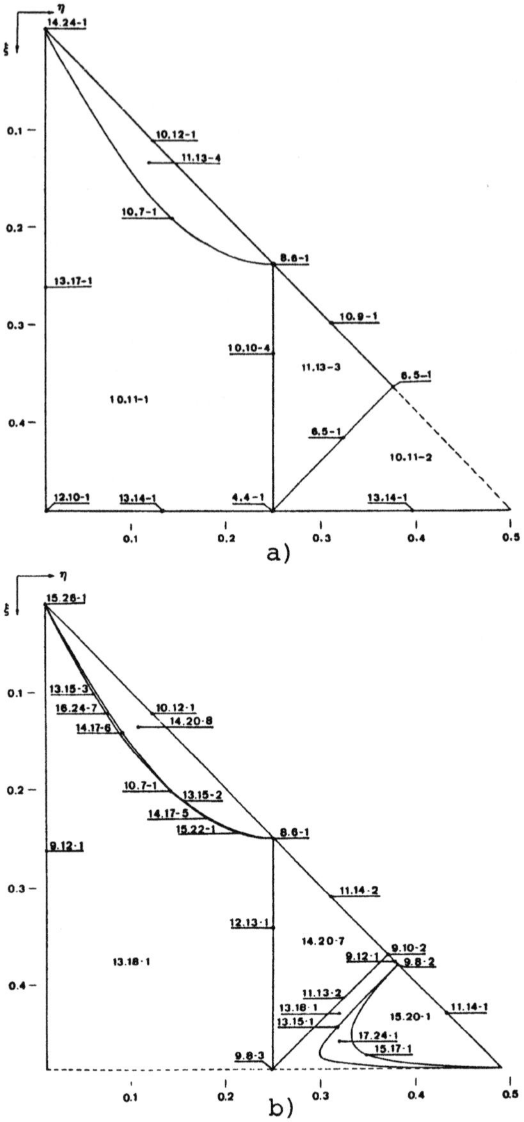

Figure 9.4. Existence charts for the isohedral types of Dirichlet domain partitions in the space group I23. a) Section at $\zeta=0$, b) section at $\zeta=0.01$

The space group I23 is symmorphic. It is a general observation that the existence charts for symmorphic space groups are relatively simple. More complicated existence charts occur for non-symmorphic space groups.

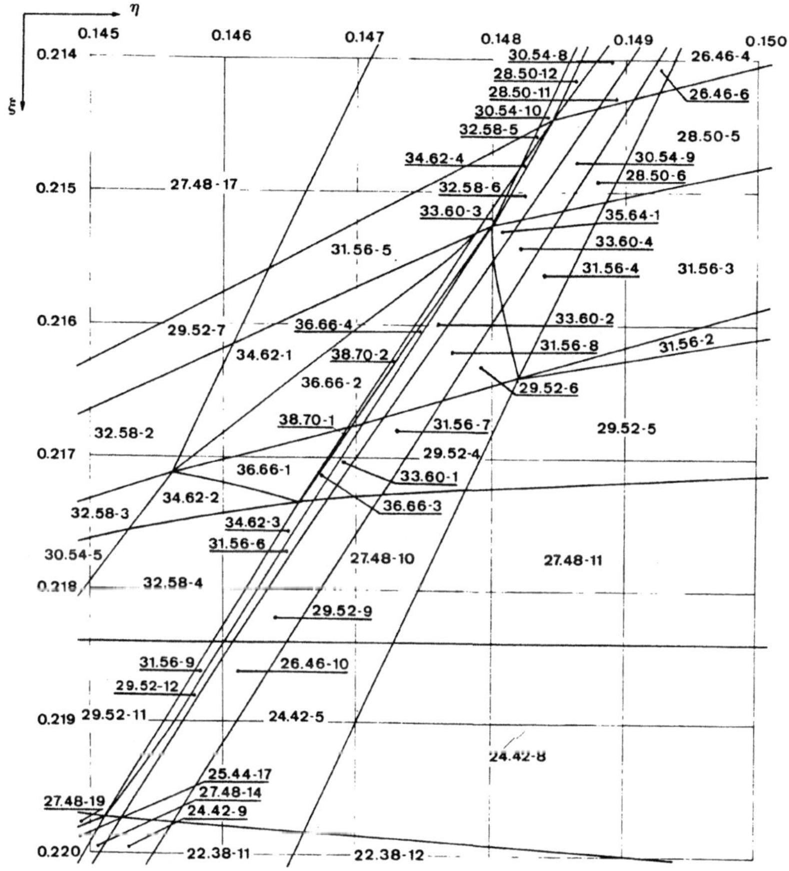

Figure 9.5. Fragment of an existence chart for the isohedral types of Dirichlet domain-partitions in the space group $I4_132$. Section at $\zeta=0.0158$

Example 9.3: Figure 9.5 shows a fragment of an existence chart for the space group $I4_132$. The parameters are within the limits $0.214\leq\xi\leq0.220$, $0.145\leq\eta\leq0.150$,

$\zeta=0.0158$. In this space group the most complicated Dirichlet domains, known at present, were discovered (Engel). Their numbers of faces amount up to 38. The Schlegel-diagrams of the four different types of Dirichlet domains with 38 faces are presented in Figure 9.6.

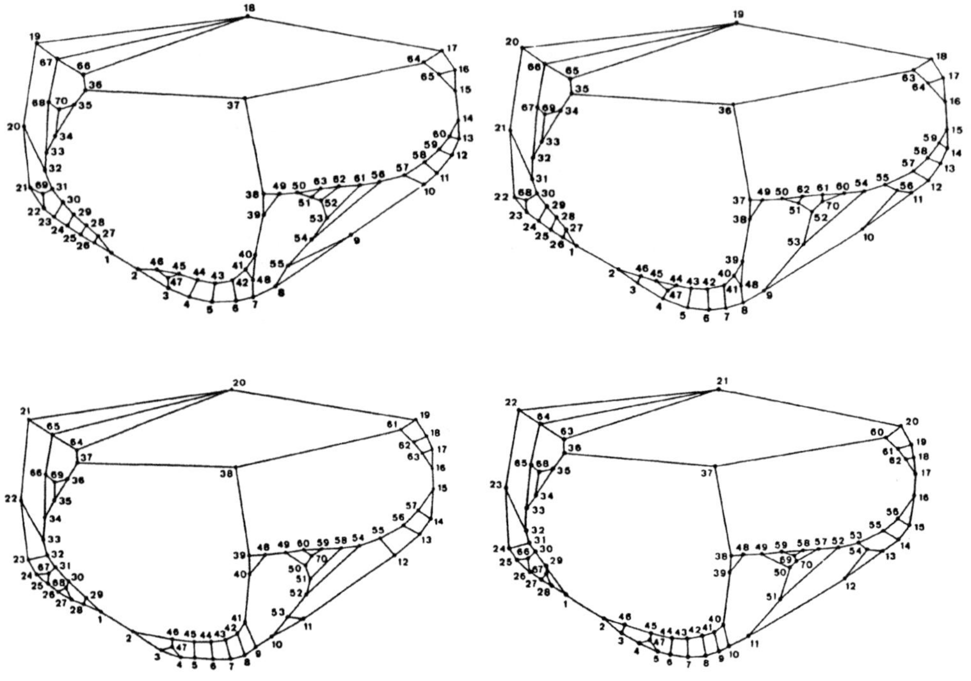

Figure 9.6. The Schlegel-diagrams of the four types of Dirichlet domains with 38 faces

Remarkably, each of these four types with 38 faces has two four-valued vertices. It is not known if these are maximal stereohedra for monohedral space-partitions.

At present a few hundred isohedral types of Dirichlet domains in E^3 are known. Considering the available results we estimate that a few thausands of different types exist.

The isohedral types of Dirichlet domain partitions correspond to the topological types of orbits which we mentioned in section 8.5.1.

9.3. Parallelotopes

General properties of Dirichlet parallelotopes were discussed in section 3.5. The concept of maximal stereotopes can again be used to derive the combinatorial types of parallelotopes in the following way:

1. The primitive parallelotopes are maximal ons. However, in dimensions $n \geq 4$ there exist non-primitive parallelotopes which are maximal. A unique characterization of the maximal parallelotopes for dimension $n > 4$ is not known.

2. All the other parallelotopes are obtained by successive zone-reductions. Let P be a parallelotope in E^n. A zone Z of P (cf. definition 3.5) is called closed if every 2-face of P that contains an edge $E_j \in Z$ also contains another edge $E_k \in Z$. A reduced parallelotope P' is obtained through the reduction of a closed zone Z of P. Let E_1 be a shortest edge of Z. Every edge $E_j \in Z$ is reduced by E_1. Thereby, the properties of a parallelotope are conserved. If P collapses under a zone-reduction then P' is a (n-1)-dimensional parallelotope.

Example 9.4: In two-dimensional space E^2 the convex centrosymmetric hexagon is the only maximal parallelogon. By a zone-reduction we obtain the centrosymmetric quadrilateral.

Table 9.3. The combinatorial types of parallelohedra in three-dimensional Euclidean space

Type	N_0	N_1	N_2	2-subordination	belts
6-1	8	12	6	4_6	4_3
8-1	12	18	8	$4_6 6_2$	$4_3 6_1$
12-1	14	24	12	4_{12}	6_4
12-2	18	28	12	$4_8 6_4$	$4_1 6_4$
14-1	24	36	14	$4_6 6_8$	6_6

In three-dimensinal space E^3 there exists one type of primitive parallelohedra, viz.: 14-1, which is a maximal one (cf. Table 9.3, and Figure 3.6). It has six equivalent zones. By a zone-reduction we get the type 12-1. The latter type has two non-equivalent zones. By reduction of the fourfold zone we obtain the type 12-2, and by reduction of the sixfold zone we get the type 8-1. From the types 12-2 and 8-1 we obtain the minimal type 6-1 by a zone-reduction. The five combinatorial types of parallelohedra were derived by Fedorov (Fedo1). A finer classification, considering their stabilizers, was performed by Delaunay (Delo2).

In four-dimensional space E^4 there exist three types of primitive parallelotopes which were determined by Voronoĭ (Voro1), viz.: 30-2, 30-3, and 30-4 (cf. Table 9.4). Necessarily these are maximal. But also the type 30-1 is maximal although it is not primitive. Remarkably it has the maximal number of facets but it has twelve vertices which are six-valent. From these four maximal types we can derive 48 reduced parallelotopes by zone-reductions as is shown in Figure 9.7. They were determined by Delaunay (Delo1) and Štogrin (Štog1).

Table 9.4. The combinatorial types of parallelotopes in four-dimensional Euclidean space

Type	N_0	N_1	N_2	N_3	2-subordination	3-subordination	belts
8-1	16	32	24	8	4_{24}	6_8	4_6
10-1	24	48	34	10	$4_{30}6_4$	6_68_4	4_76_1
12-1	36	72	48	12	$4_{36}6_{12}$	8_{12}	4_96_2
14-1	28	62	48	14	4_{48}	$6_{12}12_2$	4_66_4
14-2	36	74	52	14	$4_{44}6_8$	$6_88_412_2$	4_76_4
16-1	48	96	64	16	$4_{48}6_{16}$	$6_68_814_2$	4_76_6
20-1	30	70	60	20	4_{60}	6_{20}	6_{10}
20-2	42	94	72	20	$4_{66}6_6$	$6_{12}8_612_2$	4_36_{10}
20-3	54	114	80	20	$4_{64}6_{16}$	$6_88_812_4$	4_56_{10}
22-1	46	108	84	22	4_{84}	$6_{16}12_6$	4_36_{12}
22-2	54	116	84	22	$4_{72}6_{12}$	$6_{16}12_6$	4_36_{12}
24-1	24	96	96	24	3_{96}	8_{24}	6_{16}
24-2	30	102	96	24	$3_{72}4_{24}$	8_{24}	6_{16}
24-3	36	108	96	24	$3_{54}4_{36}5_6$	8_{24}	6_{16}
24-4	40	116	100	24	$3_{48}4_{52}$	$8_{20}10_4$	4_16_{16}
24-5	42	114	96	24	$3_{36}4_{54}6_6$	8_{24}	6_{16}

Table 9.4. (continued)

Type	N_0	N_1	N_2	N_3	2-subordination	3-subordination	belts
24-6	42	114	96	24	$3_{42}4_{36}5_{18}$	8_{24}	6_{16}
24-7	46	122	100	24	$3_{36}4_{52}5_{12}$	$8_{20}10_{4}$	$4_{1}6_{16}$
24-8	52	128	100	24	$3_{30}4_{40}5_{30}$	$8_{20}10_{4}$	$4_{1}6_{16}$
24-9	52	128	100	24	$3_{24}4_{58}5_{12}6_{6}$	$8_{20}10_{4}$	$4_{1}6_{16}$
24-10	56	136	104	24	$3_{24}4_{56}5_{24}$	$8_{16}10_{8}$	$4_{2}6_{16}$
24-11	56	136	104	24	$3_{24}4_{56}5_{24}$	$8_{18}10_{4}12_{2}$	$4_{2}6_{16}$
24-12	60	134	98	24	$4_{86}6_{12}$	$6_{10}8_{8}12_{6}$	$4_{2}6_{15}$
24-13	62	142	104	24	$3_{18}4_{50}5_{30}6_{6}$	$8_{16}10_{8}$	$4_{2}6_{16}$
24-14	62	142	104	24	$3_{12}4_{68}5_{12}6_{12}$	$8_{18}10_{4}12_{2}$	$4_{2}6_{16}$
24-15	62	142	104	24	$3_{24}4_{32}5_{48}$	$8_{18}10_{4}12_{2}$	$4_{2}6_{16}$
24-16	72	150	102	24	$4_{76}6_{26}$	$6_{8}8_{10}12_{4}14_{2}$	$4_{3}6_{15}$
24-17	72	156	108	24	$3_{24}4_{12}5_{72}$	$8_{18}12_{6}$	$4_{3}6_{16}$
24-18	72	156	108	24	$3_{12}4_{48}5_{36}6_{12}$	$8_{14}10_{8}12_{2}$	$4_{3}6_{16}$
24-19	72	156	108	24	$3_{12}4_{48}5_{36}6_{12}$	$8_{12}10_{12}$	$4_{3}6_{16}$
24-20	72	156	108	24	$4_{84}6_{24}$	$8_{16}10_{6}14_{2}$	$4_{3}6_{16}$
26-1	56	144	114	26	$3_{24}4_{90}$	$6_{2}8_{12}10_{12}$	6_{19}
26-2	62	150	114	26	$3_{18}4_{78}5_{18}$	$6_{2}8_{12}10_{12}$	6_{19}
26-3	68	156	114	26	$3_{18}4_{54}5_{42}$	$6_{2}8_{12}10_{12}$	6_{19}
26-4	68	156	114	26	$3_{12}4_{72}5_{24}6_{6}$	$6_{2}8_{12}10_{12}$	6_{19}
26-5	72	164	118	26	$3_{12}4_{70}5_{36}$	$6_{2}8_{10}10_{12}12_{2}$	$4_{1}6_{19}$
26-6	78	170	118	26	$3_{12}4_{52}5_{48}6_{6}$	$6_{2}8_{10}10_{12}12_{2}$	$4_{1}6_{19}$
26-7	78	170	118	26	$3_{6}4_{70}5_{30}6_{12}$	$6_{2}8_{10}10_{12}12_{2}$	$4_{1}6_{19}$
26-8	78	168	116	26	$4_{92}6_{24}$	$6_{8}8_{8}12_{10}$	$4_{2}6_{18}$
26-9	88	184	122	26	$3_{12}4_{38}5_{60}6_{12}$	$6_{2}8_{10}10_{8}12_{6}$	$4_{2}6_{19}$
26-10	88	184	122	26	$3_{6}4_{56}5_{42}6_{18}$	$6_{2}8_{8}10_{12}12_{4}$	$4_{2}6_{19}$
26-11	88	184	122	26	$4_{74}5_{24}6_{24}$	$6_{2}8_{8}10_{14}14_{2}$	$4_{2}6_{19}$
28-1	88	192	132	28	$3_{6}4_{72}5_{54}$	$6_{4}8_{6}10_{12}12_{6}$	6_{22}
28-2	94	198	132	28	$4_{78}5_{36}6_{18}$	$6_{4}8_{6}10_{12}12_{6}$	6_{22}
28-3	94	198	132	28	$3_{6}4_{60}5_{54}6_{12}$	$6_{4}8_{6}10_{12}12_{6}$	6_{22}
28-4	96	190	130	28	$4_{90}6_{40}$	$6_{6}8_{12}12_{6}14_{4}$	$4_{1}6_{21}$
28-5	104	212	136	28	$4_{70}5_{36}6_{30}$	$6_{4}8_{4}10_{14}12_{4}14_{2}$	$4_{1}6_{22}$
28-6	104	212	136	28	$3_{6}4_{52}5_{54}6_{24}$	$6_{4}8_{6}10_{8}12_{10}$	$4_{1}6_{22}$
30-1	102	216	144	30	$4_{108}6_{36}$	$6_{12}12_{18}$	6_{24}
30-2	120	240	150	30	$4_{90}6_{60}$	$8_{20}14_{10}$	6_{25}
30-3	120	240	150	30	$4_{72}5_{36}6_{42}$	$6_{6}8_{2}10_{12}12_{6}14_{4}$	6_{25}
30-4	120	240	150	30	$3_{6}4_{54}5_{54}6_{36}$	$6_{6}8_{6}12_{18}$	6_{25}

Only a few results are known on five-dimensional parallelotopes. According to Baranovskii and Ryškov, there exist 221 combinatorial types of primitive parallelotopes in E^5 (Bara1). Up till now we have found 133 types of primitive parallelotopes, as well as 24 non-primitive types having the maximal number of facets. We

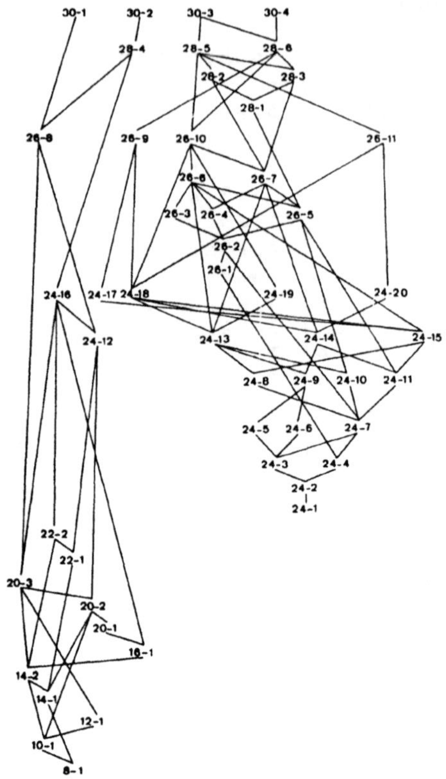

Figure 9.7. The derivation of the four-dimensional parallelotopes by successive zone-reductions

estimate the total number of parallelotopes in E^5 in the order of 10,000. Remarkably the numbers of d-faces of the primitive parallelotopes in E^n, $n \geq 5$, are not unique as can be seen from Table 9.5.

Table 9.5. The numbers of d-faces of the primitive parallelotopes in five-dimensional space

Type	N_0	N_1	N_2	N_3	N_4	belts
1	708	1770	1536	534	62	6_{89}
2	720	1800	1560	540	62	6_{90}

9.4. The regularity condition

We now consider a (r,R)-system X. From each point $x_i \in X$ we draw the straight line segments to all the other points of X. The regularity condition 1.3 requires that these line systems are directly or symmetrically congruent (cf. section 1.1). This condition can be weakened if we consider the local properties of X. In this section it will be shown that the regularity already is asserted if these line systems for every point $x_i \in X$ are directly or symmetrically congruent within a given "sphere of regularity" centered at x_i.

To every (r,R)-system corresponds a unique Dirichlet domain partition (cf. section 2.3). We will use this construction to investigate the local properties of X. However, to begin with, we consider more general space partitions and only at the end we come back to the Dirichlet domain partition. In what follows we consider a normal space partition τ with tiles T_i, $i \in I$.

The cell complex of all tiles $T_j \in \tau$ which touch the tile T_i,

$$C^1(T_i) := \{T_j \mid T_j \cap T_i \neq \emptyset\},$$

is called the first corona of T_i. Similarly,

$$C^k(T_i) := \{T_j \mid T_j \cap C^{k-1}(T_i) \neq \emptyset\},$$

is called the k-corona of T_i. We define $C^0(T_i) := T_i$. We say $C^k(T_i)$ is congruent to $C^k(T_j)$ if there exists an isometry $\varphi_{ij} \subset \mathbb{E}(n)$, φ_{ij}: $C^\nu(T_i) \rightarrow C^\nu(T_j)$, $\nu = k, k-1, \ldots, 0$.

Let $K[T_i] := \{\sigma_\nu \mid \sigma_\nu \colon T_i \rightarrow T_i\}$ be the stabilizer of T_i in the Euclidean group $\mathbb{E}(n)$. Similarly let $K[C^1(T_i)] := \{\sigma_\nu \mid \sigma_\nu \colon C^1(T_i) \rightarrow C^1(T_i),\ T_i \rightarrow T_i\}$ be the stabilizer of $C^1(T_i)$. It follows that $K[C^1(T_i)] \leq K[T_i]$ because every $\sigma \in K[C^1(T_i)]$ maps T_i onto itself. We write $K[T_i]$ as a union of cosets with respect to $K[C^1(T_i)]$,

$$K[T_i] = \hat{\sigma}_1 K[C^1(T_i)] \cup \ldots \cup \hat{\sigma}_m K[C^1(T_i)],$$

where $\hat{\sigma}_1$ is the identity operation.

Let now τ_0 be a fixed normal space partition, such that all tiles $T_i \in \tau_0$ have congruent first coronas. By this clearly τ_0 is monohedral. We take the tiles of τ_0 and try

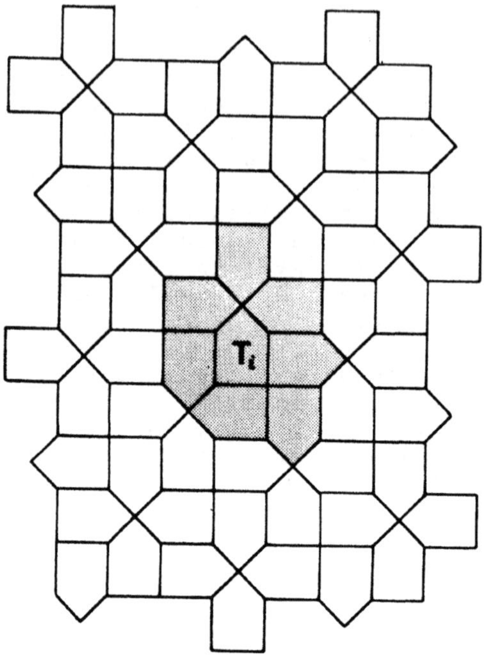

Figure 9.8. The first corona of a tile T_i

to assemble them in a different way, under the condition that each tile obtains a congruent first corona. This is like putting together a puzzle. In the following we have to formalize this game introducing the concept of geometric extension.

Definition 9.5: For $T_j \in C^1(T_i)$, $i \neq j$, an isometry $\gamma_{ij} \in \mathbb{E}(n)$, $\gamma_{ij}: T_i \to T_j$, $C^1(T_i) \to C^1(T_j)$, is called a geometric extension if the overlap domain $O_{ij} := C^1(T_i) \cap C^1(T_j)$ is a subcomplex of $C^1(T_i)$.

A geometric extension γ_{ij} requires that for $T_i \cap T_j \neq \emptyset$ it holds that $\gamma_{ij}^{-1}(T_j) = T_i$, $\gamma_{ij}^{-1}(T_i) = T_k$. Since $\gamma_{ij}^{-1}(T_i \cap T_j) = T_k \cap T_i \neq \emptyset$ it follows that $T_k \in C^1(T_i)$. Hence $\gamma_{ij}^{-1}(O_{ij}) = O_{ik} \subset C^1(T_i)$, where O_{ik} is a subcomplex of $C^1(T_i)$. The overlap domain O_{ik} may be identical to O_{ij} but, in general, this is not required.

We call an ordered sequence of tiles $T_1, \ldots, T_s$, where $T_i \cap T_{i+1} \neq \emptyset$, $i = 1, \ldots, s-1$, a chain of tiles. If $T_{s+1} = T_1$ we say that the chain is closed. We call the successive geometric extensions $\gamma_{12}, \gamma_{23}, \ldots, \gamma_{s\ 1s}$ a geometric extension along the chain $T_1, \ldots, T_s$.

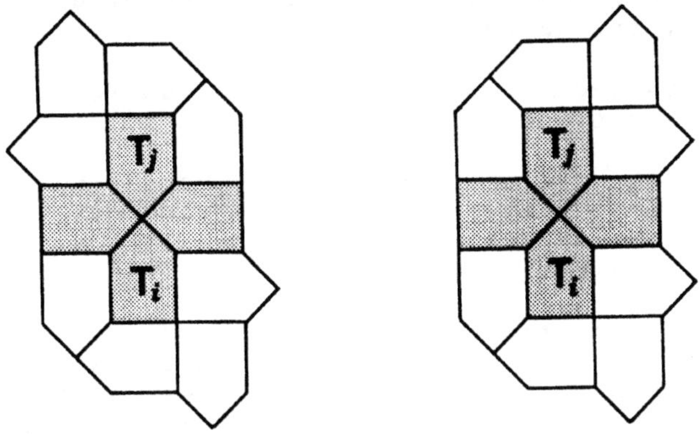

Figure 9.9. The two alternative settings for a multivaried geometric extension with overlap domain O_{ij}.

In what follows we assume that $K[C^1(T_i)]$ is a proper subgroup of $K[T_i]$. If for some λ and $\rho_v \in \hat{\sigma}_\lambda K[C^1(T_j)]$, $\rho_v : T_j \to T_j$, there exists an overlap domain $O_{jh} \subset C^1_v(T_j)$ such that $\rho_v(O_{jh}) = O_{ij}$, then for $\lambda \neq 1$,

$$\rho_v \gamma_{ij} : C^1(T_i) \overset{\gamma_{ij}}{\to} C^1(T_j) \overset{\rho_v}{\to} C^1_v(T_j)$$

is an alternative geometric extension, where $C^1_v(T_j)$ is not assumed to be a subcomplex of τ_0. Such an example is shown in Figure 9.9. In general, O_{jh} is not required to be identical to O_{ij}. For all such λ let $S_{ij} := \{\hat{\sigma}_1 K[C^1(T_j)], \ldots, \hat{\sigma}_1 K[C^1(T_j)]\}$ be the union of the corresponding cosets of $K[T_j]$. We call $\rho_v \gamma_{ij}$, $\rho_v \in S_{ij}$, a multivaried S_{ij}-extension, if S_{ij} contains more than one coset. Clearly multivaried γ_{ij}-extensions only occur if $K[C^1(T_i)] < K[T_i]$.

We next consider two geometric extensions γ_{ij} and γ_{ik}. We choose $T_k \subset C^1(T_i)$, $k \neq j$, such that $T_k \cap T_j \neq \emptyset$. Then γ_{ij} and γ_{ik} are called contiguous geometric extensions. For a given $\rho_v \gamma_{ij}$, $\rho_v \in S_{ij}$, we say $\rho_\mu \gamma_{ik}$, $\rho_\mu \in S_{ik}$, can be realized if $O_{jk} := C^1_v(T_j) \cap C^1_\mu(T_k)$ is a subcomplex of $C^1_v(T_j)$. Otherwise $\rho_\mu \gamma_{ik}$ cannot be realized. In general two contiguous multivaried extensions can only be realized for some of the cosets say $\bar{S}_{ij} \subset S_{ij}$ and $\bar{S}_{ik} \subset S_{ik}$ respectively.

We say that two contiguous multivaried geometric extensions $\rho_v\gamma_{ij}$ and $\rho_\mu\gamma_{ik}$ are not coupled if for every $\rho_v \in \bar{S}_{ij}$ the geometric extensions $\rho_\mu\gamma_{ik}$ can be realized for all $\rho_\mu \in \bar{S}_{ik}$. Otherwise they are called coupled. We have to consider this two cases separately.

1. We assume that all pairs of contiguous multivaried extensions are not coupled.

Theorem 9.6: If in a normal space partition τ_0 all tiles $T_i \in \tau_0$ have congruent first coronas and only if no coupled extensions occur then the assemblage of τ_0 is unique.

Proof: Starting from the tile $T_0 \in \tau_0$ we assemble the space partition under the condition that every tile T_i, $i \in I$, has a congruent first corona. We perform geometric extensions along all possible chains $T_0, T_j, \ldots$. Suppose we thereby encounter a multivaried S_{sv}-extension. By the congruence of the first coronas there exists a multivaried S_{0k}-extension. We prove that we can uniquely build up chains $T_0, T_1, \ldots, T_s$ such that $T_i \cap T_k \neq \emptyset$, $i \neq k$, for $i=0,1,\ldots,s$. The overlap domain $O_{0k} \subset C^1(T_0)$ is given. It contains at least a tile $T_1 \neq T_0, T_k$. We take T_1. By assumption $\rho_v\gamma_{01}$ and $\rho_\mu\gamma_{0k}$ are not coupled. Thus for every $\rho_\mu \in \bar{S}_{0k}$ we can realize all extensions $\rho_v\gamma_{01}$, $\rho_v \in \bar{S}_{01}$. Since $O_{1k} = C^1_v(T_1) \cap C^1_\mu(T_k)$ this means that O_{1k} remains conserved and hence, is uniquely determined. By the same argument as above there exists a tile $T_2 \in O_{1k}$ such that $T_2 \neq T_0, T_1, T_k$. As before we can show that O_{2k} is uniquely determined etc. In this way we construct all such chains and finally $C^1(T_k)$ becomes uniquely determined. This proves theorem 9.6.

2. We now consider the second case that coupled extensions occur. Starting from the tile $T_0 \in \tau_0$ suppose that we have made all unique extensions along all possible chains ending up with a cell complex A_0. Clearly $A_0 \cap C^1(T_0) \neq C^1(T_0)$ because otherwise, by the congruence of the first coronas, A_0 could be extended to τ_0. Let A_H be a parallel set of A_0, that is, A_H is the union of all balls $B(p,H)$ of radius H which are attached at all points $p \in A_0$. Similarly, let E_H be a parallel set of a plane E. If for a finite real costant H there exists a plane E of dimension d but not d+1 such that $A_0 \subset E_H$ and $E \subset A_H$ then we say that A_0 has an extent d, $\mathrm{ext}(A_0)=d$. We call a layer L a cell complex of extent n-1 which consists locally of a minimum number of

tiles $T_i \in \tau_0$ such that L separates E^n into two components C^- and C^+ which have distance $\vartheta>0$. For a given layer L_0 the adjacent layer L_1 is defined as

$$L_1 := \{T_v \mid T_v \in C^+,\ T_v \cap L_0 \neq \emptyset\}.$$

For a cell complex A we define the first corona as the cell complex

$$C^1(A) := \{T_v \mid T_v \cap A \neq \emptyset\}.$$

<u>Conjecture 9.7:</u> If in a normal space partition τ_0 coupled extensions occur then the assemblage of τ_0 is uniquely determined if all tiles $T_i \in \tau_0$ have congruent second coronas.

Sketch of a proof: Suppose $ext(A_0)=d<n-1$. Then there exists a plane E of dimension d and a finite radius H such that $E \subset A_H$ and $A_0 \subset E_H$. Let $E^\perp$ be the orthogonal complement of E with origin in T_0 and having dimension $d'=n-d>1$. We denote by M the cell complex

$$M := \{T_v \mid T_v \in C^1(A_0) \setminus A_0,\ T_v \cap E^\perp \neq \emptyset\}.$$

Because $d'>1$ we find in M at least one closed chain W_1 of tiles, $W_1 := \{T_1, \ldots, T_s, T_{s+1}=T_1\}$, running around A_0. For $T_1, T_2 \in W_1$ let $T_i \in A_0$ be adjacent to both T_1 and T_2. By assumption the contiguous extensions $\rho_v \gamma_{i1}$, $\rho_v \in \bar{S}_{i1}$, and $\rho_\mu \gamma_{i2}$, $\rho_\mu \in \bar{S}_{i2}$, are coupled. For a given choice of $\rho_v \gamma_{i1}$ we perform the geometric extensions along the chain W_1. Let $T_j, T_k \in W_1$ be two successive tiles of W_1, $j=1,\ldots,s$. Either for some j, $\rho_\lambda \gamma_{jk}$, $\rho_\lambda \in \bar{S}_{jk}$, is not coupled then by theorem 9.6 O_{jk} remains conserved. In this case, by the congruence of the first coronas, A_0 can be extended further. Or else $\rho_\lambda \gamma_{jk}$ is coupled, that is, the chain can be extended in at least two different ways. Thus, for $j=1,\ldots,s$, we can perform the geometric extensions along the chain W_1 in at least 2^s different ways. Since the chain W_1 is closed, we have that for each possibility the overlap domain O_{s1} must be conserved because of the given choice of $\rho_v \gamma_{i1}$. Hence, at least 2^s overlap domains of $C^1(T_0)$ have to be congruent to O_{s1}. Since there are only s overlap domains along the chain W_1 it is therefore impossible that all of them can be coupled. The same argument holds until $ext(A_0)=n-1$. The cell complex A_0 may not be a single layer because the first corona $C^1(T_0)$ determines both adjacent layers uniquely and hence τ_0 is determined uniquely. If A_0 is a double layer then the adjacent double layer is not uniquely determined because the first corona is part of exactly three layers. A_0 may not contain more than two layers because otherwise $C^1(T_0) \subset A_0$ and by the congruence of the first coronas τ_0 is uniquely determined.

The cell complex A_0 consists of a double layer. By the congruence of the second coronas, both adjacent double layers A_{-1} and A_1 are uniquely determined. Therefore, the stacking of six layers is uniquely determined. It follows that $C^2(T_0) \subset [A_{-1} \cup A_0 \cup A_1]$. By the congruence of the second coronas this is true for all tiles $T_v \in \tau_0$ hence, the stacking of all layers is uniquely determined.

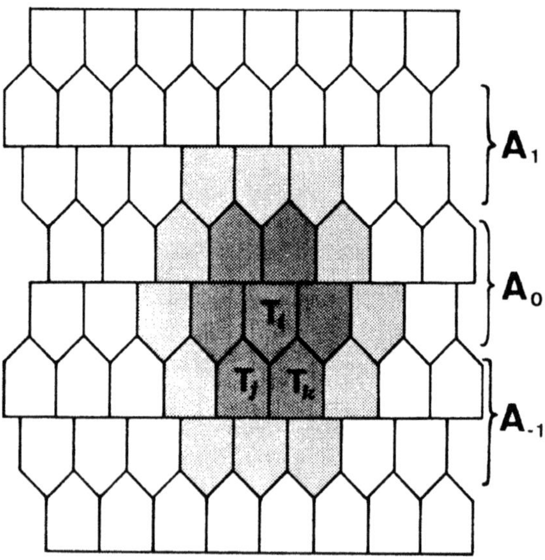

Figure 9.10. Layered structure with coupled extension

Theorem 9.8. A normal space partition τ_0 is regular if every tile of τ_0 has a congruent second corona.

Proof: By theorem 9.7 the assemblage is unique if every tile $T_v \in \tau_0$ has a congruent second corona. We have to prove that there exists an infinite group of isometries which acts transitively on the tiles of τ_0. Let the space partition τ_0 be assembled in a unique way starting from the tile $T_i \in \tau_0$. For any other tile $T_j \in \tau_0$ as starting point, we can assemble the same tiling τ_0 in a unique way. We show that there exists an isometry φ which maps T_i onto T_j and thereby maps τ_0 onto itself. Indeed, for any $T_s \in \tau_0$ we can find a chain from T_i to T_s. Because of the unique assemblage at least one congruent chain exists from T_j to some $T_v \in \tau_0$. Hence φ maps T_s onto T_v. This is true for every pair of tiles T_i and T_j and all tiles T_s. This proves theorem 9.8.

Theorem 9.9. A (r,R)-system X is regular if at every point $x_i \in X$, the system of straight line segments drawn from x_i to all other points $x_j \in X$ within a ball of radius 6R and center x_i is congruent to the corresponding system of straihgt line segments at x_0.

Proof: The (r,R)-system determines a unique Dirichlet domain partition. By theorem 2.4 and condition 1.1 it is normal. By theorem 9.8 the congruence of the second coronas is sufficient that the Dirichlet domain partition is regular, and therefore, the (r,R)-system is regular. By theorem 2.2 the Dirichlet domain of the point $x_i \in X$ is determined through all points $x_j \in X$ within a ball of radius 2R and center x_i. It follows that the first corona is determined trough all points within a ball of radius 4R and a ball of radius 6R is sufficient to determine the second corona.

The ball of radius 6R is an upper bound. In most cases a ball of radius 4R or even 2R is sufficient to ensure regularity. Since the Dirichlet domains are convex polyhedra and the Dirichlet domain partition is a very special face-to-face tiling we still have to show that the ball of radius 6R is necessary. Indeed in the plane E^2 we can easily show that no face-to-face tiling exists which requires the congruence of the second corona in order to establish the regularity. However, in three-dimensional space E^3 such face-to-face tilings exist. The first known example of a Dirichlet domain partition which requires the congruence of the second corona is shown in Figure 9.11.

The Dirichlet domain is a square prism with a slanting roof. Translates of the Dirichlet domain are assembled to form a layer. Two congruent layers are put together head-to-head to form a double layer. Congruent double layers are stacked one upon the other. Each double layer can be rotated by ±90° with respect to the previous one. Thereby the congruence of the first coronas is conserved. If we rotate always by +90° then a regular Dirichlet domain partition results with space group $P4_122$ ($P4_322$ respectively if we rotate always by -90°). If we rotate alternately by +90° and -90° then again a regular Dirichlet domain partition results which has congruent (direct or symmetric) first coronas but different second coronas with respect to the first partition. The space group is C2/c. Both space partitions belong to the same topologicl type. The second space partition can be obtained from the first one by a continuous deformation. Besides these two regular partitions an infinite number of different irregular partitions exist, where the sense of rotation is arbitrarily changed. All these partitions have congruent first coronas.

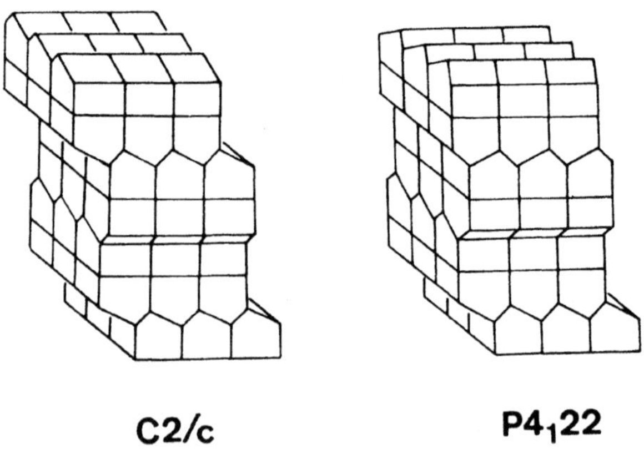

Figure 9.11. Two different regular Dirichlet domain partitions which have congruent first coronas

A list of known Dirichlet domains which show these exeptional properties is shown in Figure 9.12. In the first column is given the space group when the sense of rotation is always the same. Where two space groups are given in Figure 9.12, the rotation angle is ±120° for $P3_121$ and

$P3_112$, and ±60° for $P6_122$. These correspond to different partitions. The second column shows the Dirichlet domain and the third column gives a view from top of it.

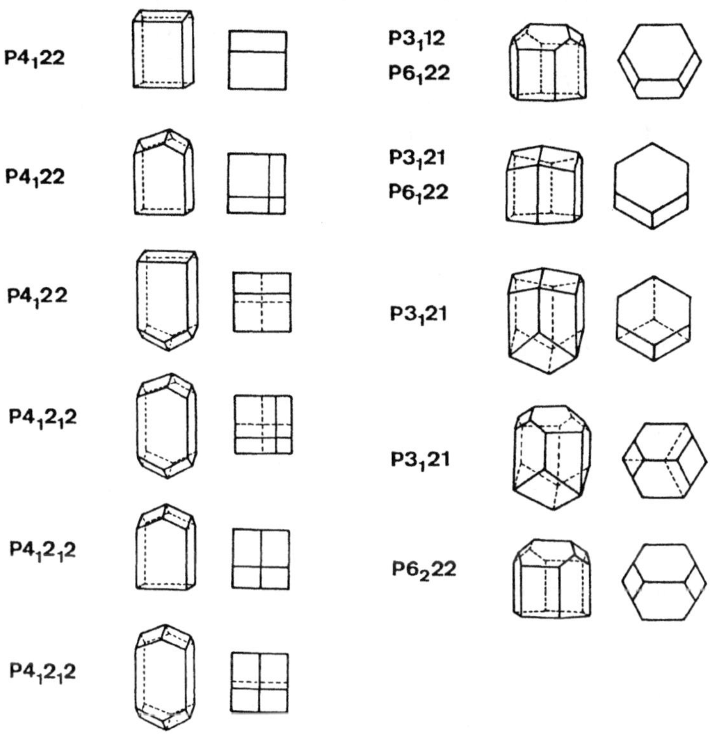

Figure 9.12. List of known Dirichlet domains which allow irregular space partitions each having congruent first coronas

At present these are the only known examples of point sets in E^3 which have a very high local regularity but which may be globally irregular. These kinds of point sets are fundamental in order to understand the crystalline state of matter.

9.5. Dissections of polytopes

We consider closed polytopes in Euclidean space E^n. A polytope P is said to be dissected into P_1 and P_2 if $P=P_1 \cup P_2$ and the interiors of P_1 and P_2 are disjoint. Following Sah (Sah1), we will write $P_1 \coprod P_2$ to denote the pairwise interior disjoint union of P_1 and P_2. A polytope P_1 is said to be congruent to the polytope P_2 if there exists an isometry $\varphi \in \mathbb{E}(n)$, such that $P_2 = \varphi P_1$. We define the volume of P through the additive function $f_0(P):=\mathrm{vol}(P)$ having values in $\mathbb{R}$ (set of real numbers) such that the following conditions hold:

1. $f_0(P_1)=f_0(P_2)$, if P_1 is congruent to P_2,
2. $f_0(P_1 \coprod P_2)=f_0(P_1)+f_0(P_2)$,
3. $f_0(P) \geq 0$.

In the theory of volume the equidissection is of importance.

Definition 9.6: Two polytopes P and Q are called equidissectable, $P \sim Q$, if P and Q can be finitely dissected into $P_1,\ldots,P_s$, and $Q_1,\ldots,Q_s$ respectively, such that P_i is congruent to Q_i, for $i=1,\ldots,s$.

Necessarily equidissectable polytopes have the same volume.

From elementary geometry in the plane E^2 it is known that two polygons P and Q are always equidissectable if they have equal areas. However, in spaces of higher dimensions, in general, this is no longer true. This means that the theory of volume depends upon the axiom of continuity. In answering Hilbert's third problem, Dehn gave necessary conditions for two polyhedra in E^3 to be equidissectable (Dehn1).

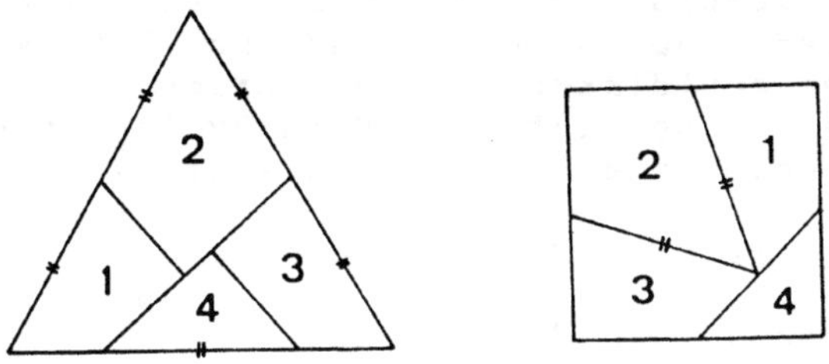

Figure 9.13. Dissection of an equilateral triangle into a square of equal area

Let P be a polytope in E^n, $n\geq 3$. Following Dehn we determine the function

$$f_1(P) = \sum_{i=1}^{N} vol(F_i)\alpha_i(F_i),$$

where the summation extends over the N (n-2)-faces F_i of P. Here $vol(F_i)$ is the (n-2)-dimensional volume of F_i and $\alpha_i(F_i)$ is the interior dihedral angle, measured in radians, between both adjacent facets which meet in F_i. Nowadays the first Dehn function is expressed as a tensor product in the following way:

Every real number $\alpha_i \in \mathbb{R}$ can be represented by a finite linear combination,

$$\alpha_i = \xi_{i1}\pi + \xi_{i2}\gamma_2 + \ldots + \xi_{ik}\gamma_k,$$

where $\xi_{ij} \in \mathbb{Q}$ (set of rational numbers), and $\gamma_1 := \pi$, $\gamma_2, \ldots, \gamma_k$ are rationally independent basis vectors of a $\mathbb{Q}$-vector space. We define $\mathbb{R}/Q\pi := \mathbb{R} \bmod (\mathbb{Q}\pi)$. The first Dehn function thus becomes,

$$f_1(P) = \sum_{i=1}^{N} vol(F_i) \times \alpha_i(F_i).$$

It has values in $\mathbb{R} \times \mathbb{R}/\mathbb{Q}\pi$. We now consider a certain finite dissection $P = \coprod P_i$. The first Dehn function,

$$f_1(P) = \sum_{i=1}^{N} f_1(P_i).$$

is invariant (modulo $\mathbb{Q}\pi$) of the particular dissection chosen. Sydler and Jessen proved that Dehn's function is sufficient for equidissection in three- and four-dimensional space. We state the Dehn-Sydler theorem without proof (for a proof see Sydl1, Jess1, Sah1).

Theorem 9.10: Let P and Q be two polytopes in Euclidean space E^n, $n \le 4$. If P and Q agree in their Dehn functions f_0 and f_1 (modulo $\mathbb{Q}\pi$), then they are equidissectable.

In spaces of dimension $n>4$ necessarily higher order Dehn's functions have to be considered. Yet there is still no proof that such conditions are sufficient.

The importance of the Dehn-Sydler theorem is revealed in the following theorem, due to Debrunner, which we state without proof (for a proof see Debr2).

Theorem 9.11: Let τ be a normal space partition in E^n with stereotopes P_ν, $\nu \in I$, all of which (except of a finite number) are similar to P_0, $P_\nu = \varphi_\nu \lambda_\nu P_0$, $\varphi_\nu \in \mathbb{E}(n)$, $0<\lambda_\nu \in \mathbb{R}<\infty$. Then P_0 has first Dehn function $f_1(P_0)=0 \bmod(\mathbb{Q}\pi)$.

The hypercube H_n in E^n has first Dehn function $f_1(H_n)=0$. Under the condition of theorem 9.11 it follows from theorem 9.10 that every stereotope in E^n, $n \le 4$, is equidissectable to a hypercube H_n of equal volume.

In the following we state some results on dissection relations between regular simplices A_n and crosspolytopes C_n in E^n. The dihedral angles of these regular polytopes are given by Coxeter (Coxe1) as follows:

$$A_n:\ \cos\alpha_n = 1/n, \qquad \mathrm{tg}\,\alpha_n = \sqrt{n+1}\sqrt{n-1}$$
$$C_n:\ \cos\beta_n = (2-n)/n, \quad \mathrm{tg}\,\beta_n = 2\sqrt{n-1}/(2-n).$$

The dihedral angles α_n and β_n, measured in radians, are linearly dependent in the $\mathbb{Q}$-vector space $\mathbb{R}/\mathbb{Q}\pi$. For certain integers p,q,r there exists a relation,

$$p\alpha_n \pm q\beta_n = \pm r\pi,$$

which provides the possible dissection relations. The following theorem, due to Debrunner, we state without proof (for a proof see Debr1).

Theorem 9.12: A dissection relation between regular simplices A_n and crosspolytopes C_n of the form

$$xf_1(A_n) + yf_1(C_n) = 0 \bmod(\mathbb{Q}\pi)$$

only exists in the following dimensions $n\geq 3$:

$$\begin{aligned} E^3:&\quad 2f_1(A_3) + f_1(C_3) = 0 \bmod(\mathbb{Q}\pi),\\ E^4:&\quad f_1(C_4) = 0 \bmod(\mathbb{Q}\pi),\\ E^8:&\quad 1920f_1(A_8) + 135f_1(C_8) = 0 \bmod(\mathbb{Q}\pi). \end{aligned}$$

The plane E^2 can be tessellated by congruent copies of an equilateral triangle. For simplices A_n in spaces of dimensions $n\geq 3$ the first Dehn function is not zero therefore, by theorem 9.11, A_n is not a stereohedron for any $n\geq 3$.

9.6. Historical remarks

Tessellations of the plane by polygons are known since the antique. The ancient Greeks associated a Platonic solid to each of their four elements, viz.: fire, earth, air, water, and they asked wether it is possible to fill space completely by congruent copies of these fundamental particals. Aristotle taught that two of the five Platonic solids, the cube and, incorrectly, the tetrahedron, are space fillers. He thus rejected the other Platonic solids as fundamental particles because in his opinion a vacuum does not exist in nature. Since these early beginnings, the problem of filling space with polyhedra had occupied many scientists up to our time, and still there are many open and difficult problems.

Particularly, in crystallography space partitions always were considered important for the description of crystals and their atomic structure. Domenico Guglielmini (1655-1710) proposed four polyhedral particals, viz.: the cube, the hexagonal prism, the rhombohedron, and the octahedron which combine to form the crystals. For octahedral crystals he assumed tetrahedral interstices in order to get a complete space filling. The parallelepipedal building bricks of Haüy were already mentioned in section 1.6.

By the beginning of the 19th century the lattice hypotheses was established by Seeber, Frankenheim, and Bravais, and in 1850 Dirichlet introduced his important construction of the Dirichlet parallelogon (cf. section 4.6). The concept of space filling polyhedra was again taken up by Fedorov who discovered in 1885 the five combinatorial types of parallelohedra in three-dimensional space. According to Fedorov a crystal consits of identical polyhedral particals which fill space completely when arranged in parallel position (Fedo5).

Parallelotopes in higher dimensional spaces were investigated in 1908 by Voronoĭ. He also found the four-dimensional primitive parallelotopes (Voro1). These results were taken up in 1929 by Delaunay who determined 51 combinatorial types of four-dimensional parallelotopes. The missing one was discovered in 1973 by M.I. Štogrin (Štog1).

Only after Max von Laue's (1879-1960) discovery of the diffraction of X-rays by crystals in 1912 the lattice hypotheses was proved and soon afterwards a lot of structural information became available. Based on these new results the crystallographer Paul Niggli in 1927 introduced the topological structure analysis. He investigated the topological connectivity by constructing aroud each atomic position its Dirichlet domain which he called the domain of influence (Nigg5). This idea of Niggli was taken up by the crystallographer Fritz Laves (1906-1978) who determined in 1931 the topological types of Dirichlet domain partitions in the plane (Lave1,Lave2). More general results on plane tessellations were already obtained by the mathematician Karl Reinhardt in 1918 (Rein1). Dirichlet domains in three-dimensional space were constructed by Nowacki in 1935 (Nowa3). Due to the limited accuracy he attained at that time it was not possible to determine the more complicated Dirichlet domains. This could only be achieved in recent years by the use of computers as it was done by H.-D. Löckenhoff and Erwin Hellner (Löck1), Elke Koch (Koch1), Peter Engel (Enge1,Enge2), and Zbigniew Dauter (Daut1).

The regularity condition of a regular point system is of fundamental importance in order to understand the crystalline state of matter. In 1888 Schönflies stated, without proof, that a monohedral space partition is regular (isohedral), if all its tiles are surrounded by its "neigbouring" tiles in a congruent way (Schö2). In general, this statement is false. First results concerning the local properties of regular point systems were obtained in 1976 by B.N. Delaunay and his colleagues (Delo3). In the endeavour to improve their upper bound, $(\nu+2)2R$, Galiulin incorrectly stated in 1980 that 4R is a sufficient radius for a sphere of regularity (Gali3). The Dirichlet domain partitions which require a regularity radius of 6R were recently discovered by the author.

According to Euclid, the volumes of triangular pyramids of equal altitude are to each other as the areas of their bases. Hilbert did not believe that the theory of volume can be based on dissection of polytopes and in his third problem he sought for two terahedra of equal altitude and equal base area which are not equidissectable (Hilb1) (cf. section 8.8). Such a counterexample and the necessary conditions were established by the mathematician Max Wilhelm Dehn (1878-1952). The theory of volume was further developed by the mathematician Hugo Hadwiger (1908-1981) (Hadw1). In 1965 the mathematician Jean-Pierre Sydler could prove that Dehn's functions are also sufficient for the equidissection of polyhedra. The important property that every prototile of a monohedral normal space partition has vanishing first Dehn function was proved by the mathematician Helmut Groemer in 1964 (Groe1) and more generally by the mathematician Hans Debrunner in 1980 (Debr2).

10. Packings of Balls

Atoms may be considered as small rigid balls. The packing of balls is an important concept in crystallography for the description of crystal structures, that is, the arrangement of atoms in a crystal. Many inorganic crystal structures are based on densest packings of balls.

10.1. <u>Definitions</u>

In Euclidean space E^n we consider a family of n-dimensional closed balls,

$$\kappa := \{clB(x_\nu, \mu_\nu) \mid \nu \in I\},$$

where x_ν and μ_ν are the center and the radius of the closed ball clB, and I is an infinite set of indices. If two balls $clB(x_i, \mu_i)$ and $clB(x_j, \mu_j)$ touch each other then we join their centers with a straight line segment.

<u>Definition 10.1:</u> A family of balls κ is said to form a packing into E^n

1. if the intersection of the interiors of any two balls of κ is empty,

$$intB(x_i, \mu_i) \cap intB(x_j, \mu_j) = \emptyset, \quad i \neq j.$$

2. if there exists for any two balls of κ a chain of joins connecting them.

A packing of balls into E^n is called <u>monospherical</u> if all of its balls have the same finite radius μ. It is called <u>regular</u> or <u>homogeneous</u> if there exists a space group G which acts transitively on the balls. A regular packing of balls necessarily is monospherical. It is called a <u>lattice packing of balls</u> if the translation subgroup T^n of G acts transitively on the balls. A regular packing of balls is called <u>stable</u> if no ball can be moved without disturbing some other balls. The set of all centers x_ν, $\nu \in I$, together with the joins form a <u>packing graph</u> showing the topological properties of the packing.

In what follows we only consider regular packings of balls. The set of all centers x_v form a regular point system X in E^n. Clearly, by the condition of regularity, the packing graph is vertex-transitive. By $D(x_v)$ we denote the Dirichlet domain of the center x_v with respect to X (cf. section 2.1). It holds that $B(x_v,\mu)\subset \mathrm{cl}D(x_v)$. The density $\rho(\kappa)$ of a regular packing of balls is defined by

$$\rho(\kappa) := \mathrm{vol}(B(x_v,\mu))/\mathrm{vol}(D(x_v)).$$

It can be regarded as the proportion of the whole of space covered by the family of balls. The volume of a n-dimensional ball of radius μ is given by

$$\mathrm{vol}(B) = \frac{\pi^{\frac{1}{2}n}}{\Gamma(\frac{1}{2}n+1)}\,\mu^n,$$

where the gamma function is given by $\Gamma(1)=1$, $\Gamma(k+1)=k!$, $\Gamma(\frac{1}{2})=\sqrt{\pi}$, and $\Gamma(\frac{1}{2}+k)=(\frac{1}{2}+k-1)(\frac{1}{2}+k-2)\ldots(\frac{1}{2})\sqrt{\pi}$, $k\in\mathbb{N}$.

Definition 10.2: Two regular packings of balls κ and κ' are isomorphic and belong to the same topological type of regular packing if they have isomorphic packing graphs.

It may be very difficult to verify the isomorphism of two packing graphs. No algorithm which allows a unique classification of packing graphs has been described hitherto. A possible way to do it would be to look at certain polyhedral cells P_i determined by the packing graph. Since the packing graph is vertex-transitive it is sufficient to consider those cells only which meet in a vertex x_v. Generally, these cells are not convex. Only for n=2 the cells P_i are always bounded polygons because of the packing conditions. In this case it is sufficient to specify in cyclic order the s polygons which meet in the vertex x_v in the following way:

$$p_1-p_2-\ldots-p_s,$$

where p_i, $i=1,\ldots,s$, gives the number of vertices of the polygon P_i. In order to make this scheme unique we require that p_1 is maximal. Among the remaining schemes we retain those with maximal p_2 etc.

A finer classification is obtained if we consider the stabilizer G_s of the regular packing of balls and the way G_s acts on the balls.

Definition 10.3: Two regular packings of balls κ and κ' are isomorphic and belong to the same type of regular packing

1. if κ and κ' belong to the same topological type of regular packing.

2. if κ and κ' have equivalent stabilizers G_s and G'_s respectively, that is, if there exists an affine mapping φ such that $G'_s=\varphi G_s \varphi^{-1}$.

3. if the balls B_v and B'_v have equivalent stabilizers $St(B_v)$ and $St(B'_v)$ in G_s and G'_s respectively such that $St(B'_v)=\varphi St(B_v)\varphi^{-1}$.

Following Niggli, the conditions for the existence of a regular packing κ of balls are determined as follows (Nigg5): Let G be the generating space group of κ. We consider the set of all crystallographic orbits $O(G,x_0)$ of G (cf. section 8.1). Let H be a certain cyclic subgroup of G. With respect to H each crystallographic orbit $O(G,x_0)$ can be devided into two disjoint subsets,

$$O(G,x_0) = O(H,x_0) \cup O'.$$

There exists two points $x_1 \in O(H,x_0)$ and $x_2 \in O'$ such that $|\vec{x}_1-\vec{x}_0|$ and $|\vec{x}_2-\vec{x}_0|$ are minimal.

Definition 10.4: The symmetry domain $\Delta(H) \subset E^n$ of a cyclic subgroup $H<G$ is the set of all points x_0 for which

$$\inf_{x_i \in O(H,x_0)} |\vec{x}_i-\vec{x}_0| < \inf_{x_j \in O'} |\vec{x}_j-\vec{x}_0|,$$

We note that the symmetry domain $\Delta(H)$ not necessarily is bounded. For some cyclic subgroups $H'<G$ it also may be empty. For a given space group G with metric tensor C the space E^n is uniquely partitioned into symmetry domains. The boundary regions between two adjacent symmetry domains are obtained by the condition

$$\inf_{x_i \in O(H,x_0)} |\vec{x}_i-\vec{x}_0| = \inf_{x_j \in O'} |\vec{x}_j-\vec{x}_0|.$$

Suppose that the infimum of $|\vec{x}_i-\vec{x}_0|$, $x_i \in O(H,x_0)$, is attained for $\vec{x}_1 := S_1\vec{x}_0+\vec{s}_1$, $(S_1,\vec{s}_1) \in G$. We set $\vec{d}_1 := S_1\vec{x}_0+\vec{s}_1-\vec{x}_0$. It follows that

$$|\vec{x}_1-\vec{x}_0|^2 = \vec{d}_1^t C \vec{d}_1$$

(cf. section 1.2). Thus the boundary regions are surfaces of the second degree which depend both on the space group G and on the metric tensor C. We note that the invers symmetry operation $(S_1,\vec{s}_1)^{-1}$ gives the same infimum. In order to obtain a regular packing of balls into E^n at least n shortest distances $|\vec{x}_i-\vec{x}_0|$ of equal length must exist. Following Nowacki (Nowa3), it is necessary and sufficient that the corresponding n symmetry operations $(S_i,\vec{s}_i)$, i=1,...,n, form a generating set of G (cf. definition 6.4).

10.2. Packing of disks into E^2

The types of regular packings of disks into the plane E^2 were determined by Niggli (Nigg5). The results of these calculations are summerized in the following theorem.

Theorem 10.1: There exist 31 types of regular packings of disks into the plane E^2.

From the 31 types of regular packings of disks we can deduce the 11 topological types of regular packings of disks which are shown in Figure 10.1. The type 3-3-3-3-3-3 (cf. Figure 10.1a) is a densest lattice packing of disks with density $\rho=\pi/\sqrt{12}$. Its stabilizer is p6mm. It was proved by Tóth (Tóth1) that this is the closest packing of disks into E^2. The type 12-12-12 (cf. Figure 10.1l) has the lowest density, $\rho\sim 0.3907$. Its stabilizer is also p6mm.

Nowacki investigated regular packings of ellipses into E^2 (Nowa6). His list was corrected and extended by Grünbaum and Shephard (Grün3). The result is summerized in the following theorem.

Theorem 10.2: There exist 57 types of regular packings of ellipses into the plane E^2.

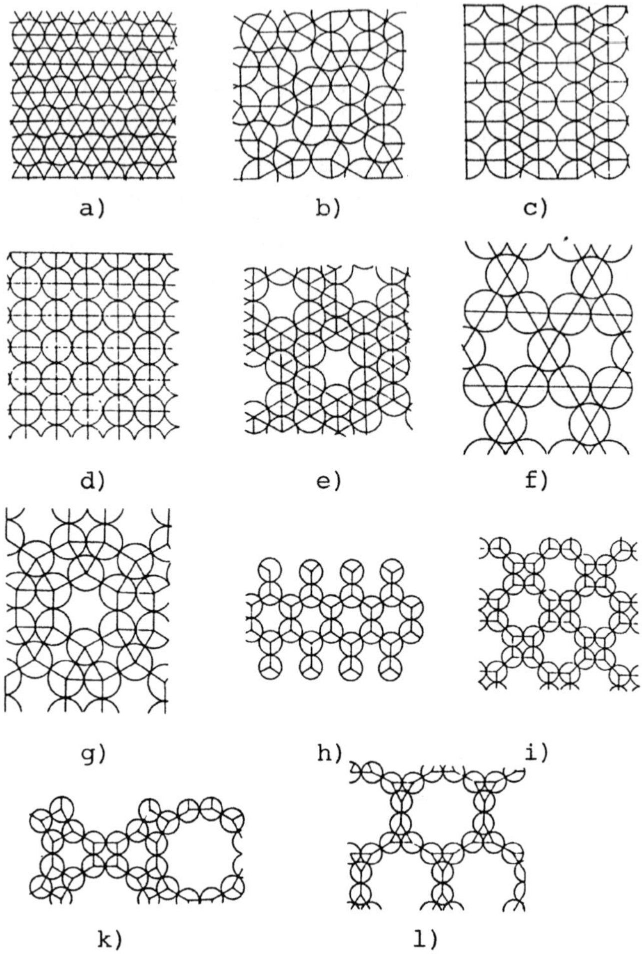

Figure 10.1. The 11 topological types of regular packings of disks into the plane E^2.

10.3. Packing of balls into E^3

Dense packings of balls into E^3 are particularly important in crystallography. We consider in E^3 a plane with a densest packing of disks and replace each disk through a ball thus obtaining a densest layer of balls. The densest lattice packing of balls is generated by stacking such layers as closely as possible, that is, the second layer fits into the voids of the first layer etc. We denote by $\vec{v}$ the stacking vector which brings the first layer A into the second layer B. There exist two possible stacking vectors $\vec{v}_1$ and $\vec{v}_2$. By applying multiples of $\vec{v}_1$ (or $\vec{v}_2$) the stacking sequence ...ABCABC... results. The lattice packing thus obtained is called the cubic closest packing. It corresponds to the cubic face-centered lattice with stabilizer $Fm\bar{3}m$. Other stacking sequences do not give lattice packings. Applying alternately $\vec{v}_1$ and $\vec{v}_2$ the stacking sequence ...ABAB... results. The packing thus obtained is regular and is called the hexagonal closest packing. Its stabilizer is $P6_3/mmc$. All other stacking sequences are not regular. The density of such packings by stacking densest layers of balls is $\rho=\pi/\sqrt{18}$. There is no proof that this is the densest possible packing of balls into E^3.

The numbers of different stacking sequences having a given repeat period were determined by Iglesias and by McLarnan (Igle1, McLa1).

The packings of balls by stacking densest layers have two different voids. The larger void is called the octahedral void. It can obtain a ball of radius $\mu(\sqrt{2}-1)$ which will be in contact with six balls situated at the vertices of an octahedron. The smaller void is called the tetrahedral void which is coordinated by four balls situated at the vertices of a regular tetrahedron. A ball of radius $\mu(\sqrt{3/2}-1)$ can be fitted into this void. There exist twice as many tetrahedra than octahedra. Tetrahedra and octahedra together completely fill the space. In the cubic closest packing the octahedra only share common edges and vertices whereas, in the hexagonal closest packing the octahedra form columns along the c-axis by sharing common faces as can be seen in Figure 10.2. Two adjacent columns only share common edges and vertices.

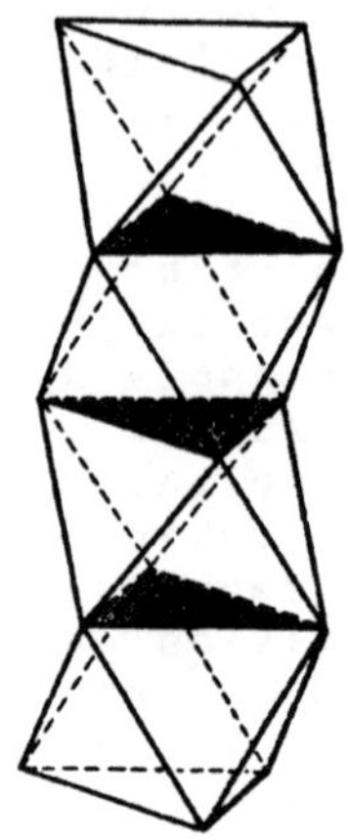

Figure 10.2. Columns of octahedra in the hexagonal closest packing

Regular packings of balls into E^3 were investigated by Sinogowitz who found 681 types of regular packings of balls generated by space groups belonging to the triclinic, monoclinic, and orthorhombic crystal systems (Sino1). Fischer determined 593 types of regular packings generated by space groups of the tetragonal and cubic crystal system (Fisc1).

Regular packings of balls of low density were investigated too. The packing of lowest density was discovered by Heesch and Laves in the space group $I4_132$ (Hee1, Fisc2).

Regular packings of ellipsoids into E^3 were investigated by Matsumoto and Nowacki (Mats1).

10.4. Lattice packings of balls in E^n

In spaces of higher dimensions mainly lattice packings of balls were investigated. The stacking process described in section 10.3 can be used to generate lattice packings of balls in any dimension (Leec1, Conw1). Lattice packings of balls were determined up to dimensions 60. However, little is known on general regular packings of balls. It was proposed by Rogers that probably for n≥5 the densest packings of balls are not lattice packings (Roge1). Leech and Sloane have found in E^{10} and E^{11} packings of balls which are denser than the densest lattice packing (Leec1). An upper bound δ for the density of packings of balls into E^n was given by Rogers in the following way (Roge1): In E^n a system of (n+1) balls are densest packed if they are situated at the vertices of a regular simplex A_n in E^n. Let σ_n denote the ratio of the volume of the part of the simplex covered by the balls to the volume of the whole simplex. For n≤2 congruent copies of a regular simplex can tile space. It follows that $\delta=\sigma_n$ for n≤2. By theorem 9.12, regular simplices in spaces of dimensions n≥3 are not space fillers. Thus $\delta<\sigma_n$ for n≥3. In Table 10.2 are listed the known densest packings of balls into E^n up to dimension n=12 according to Leech and Sloane (Leec1).

Table 10.1. Dense packings of balls into E^n

n	type	density		contact number	
			upper bound		upper bound
1	l*	1.0	1.0	2	2
2	l	0.9069	0.9069	6	6
3	l	0.7404	0.7796	12	12
4	l	0.6168	0.6478	24	26
5	l	0.4652	0.5257	40	48
6	l	0.3729	0.4192	72	85
7	l	0.2953	0.3298	126	146
8	l	0.2537	0.2567	240	244
9	l	0.1458	0.1981	272	401
10	l	0.0920	0.1518	336	648
	g	0.0946		≤372	
11	l	0.0604	0.1156	432	1035
	g	0.0662		≤566	
12	l	0.0494	0.0876	648	1637

*) l: lattice packing of balls
g: general packing of balls

Remarkably, in E^8 the densest lattice packing of balls has a density very close to the upper bound. This is because by theorem 9.12, there exists a dissection relation between the regular simplices A_n and the crosspolytopes C_n.

10.5. Historical remarks

The first treatise on packings of balls is considered to be Johann Kepler's remarkable speculation on the forms displayed by snowflakes in 1611. Kepler discussed simple cubic packing, face and body centered cubic packings and simple hexagonal packing. Robert Hook (1635-1703), however, seems to have been the first to consider the spherical particals to be atoms. He proposed that all crystalline forms can be explained by a few basic packing arrangements of balls. Similarly, Christian Huygens (1629-1695) postulated ellipsoidal atoms to explain the cleavage of calcite. Packings of balls were futher investigated by William Barlow (1845-1934) who listed five packing types; one of these was the hexagonal closest packing, which he was the first to introduce.

A systematic investigation of packings of balls was started by the crystallographer Paul Niggli in his treatise on topological structure analysis. In 1927 he derived the 31 types of regular packings of disks. Extending the ideas of Niggli, Ulrich Sinogowitz determined in 1944 packings of balls in three-dimensional space for space groups of low symmetry. In a series of papers, 1970-1975, Werner Fischer determined the types of regular packings of balls for space groups of tetragonal and cubic symmetry.

Packing of ellipsoids were again investigated by Werner Nowacki in 1948.

The mathematical theory of packing goes back to Ludwig August Seeber (Seeb2) and to Carl Friedrich Gauss (Gaus1). Important contributions to the theory of lattice packings of balls were achieved by Hermann Minkowski in 1905 giving an upper bound for the density and by H. F. Blichfeldt in 1934 who found the densest lattice packing of balls in E^8. The best upper bound for the density of packings of balls is due to Claude Ambrose Rogers. Lattice packings of balls were futher investigated by John Leech in a series of papers from 1964-1971.

References

Alex1 Alexander, E., Herrmann K.: Die 80 zweidimensionalen Raumgruppen. Z. Kristallogr. 70 (1929) 309-327

Asch1 Ascher, E., Janner, A.: Algebraic aspects of crystallography. I. Space groups as extensions. Helv. Phys. Acta 38 (1965) 551-572

Asch2 Ascher, E., Janner, A.: Algebraic aspects of crystallography. II. Non-primitive translations in space groups. Commun. Math. Phys. 11 (1968/69) 138-167

Bara1 Baranovskii, E.P., Ryškov, S.S.: Primitive five-dimensional parallelohedra. Soviet Math. Dokl. 14 (1973) 1391-1395

Barl1 Barlow, W.: Ueber die geometrischen Eigenschaften homogener starrer Structuren und ihre Anwendung auf Krystalle. Z. Kristallogr. 23 (1894) 1-63

Belo1 Belov, N.V., Belova, E.N.: Color symmetry groups. Sov. Phys. Crystallogr. 1 (1956) 5-11

Belo2 Belov, N.V., Kuntsevich, T.S.: Four-dimensional Bravais lattices. Acta Crystallogr. A25 (1969) 53

Bieb1 Bieberbach, L.: Ueber die Bewegungsgruppen der euklidischen Räume. Erste Abhandlung. Math. Ann. 70 (1910), 297-336

Bieb2 Bieberbach, L.: Ueber die Bewegungsgruppen der euklidischen Räume. Zweite Abhandlung. Die Gruppen mit endlichem Fundamentalbereich. Math. Ann. 72 (1912), 400-412

Bieb3 Bieberbach, L., Schur, I.: Ueber die Minkowskische Reduktionstheorie Sitzungsber. kgl. Preuss. Akad. der Wiss. in Berlin. Phys. Math. Klasse (1928) 510-535 Erratum: ibid. (1929) 508

Bill1 Billiet, Y.,Burzlaff, H. and Zimmermann, H.: Comments on the paper of H. Burzlaff and H. Zimmermann "On the choice of origin in the description of space groups". Z. Kristallogr. 160 (1982) 155-157

Blic1 Blichfeldt, H.F.: A further reduction of the known maximum limit to the least value of quadratic forms. Bull. Amer. Math. Soc. 23 (1917) 401

Blic2 Blichfeldt, H.F.: The minimum values of positive quadratic forms in six, seven and eight variables. Math. Zeitschrift 39 (1934) 1-15

Bohm1 Bohm, J.: Zur Anzahl kristallographischer Symmetriegruppenarten. Z. Kristallogr. 150 (1979) 115-123

Brad1 Bradley, C.J., Cracknell, A.P.: The mathematical theory of symmetry in solids. Representation theory for point groups and space groups. Clarendon press, Oxford (1972)

Brav1 Bravais, A.: Mémoire sur les polyèdres de forme symétriques. J. Math. de M. Liouville 14 (1849) 141

Brav2 Bravais, A.: Mémoire sur les systèmes formés par des points distribués regulièrement sur un plan ou dans l'espace. J. de l'école Polytechn. 19 (1850) 1-128.

English translation by A. L. Shaler. The Crystallographic Soc. Am. (1949)

Bron1 Brondsted, A.: An introduction to convex polytopes. Springer, New York (1983)

Brow1 Brown, H., Bülow, R., Neubüser, J., Wondratschek, H. and Zassenhaus, H.: Crystallographic groups of four-dimensional space. John Wiley, New York, (1978)

Buer1 Buerger, M.J.: Introduction to crystal geometry. Mc Graw-Hill Inc. New York, (1971)

Bülo1 Bülow, R.: Eine Ableitung der Kristallklassen im R^4 mit Hilfe gruppentheoretischer Programme. Diplomarbeit, Univ. Kiel, (1967)

Bülo2 Bülow, R., Neubüser, J. and Wondratschek, H.: On crystallography in higher dimensions. II. Procedure of computation in R^4. Acta Crystallogr. A27 (1971) 520-523

Burc1 Burckhardt, J.J.: Zur Theorie der Bewegungsgruppen. Comm. Math. Helv. 6 (1934) 159-184

Burc2 Burckhardt, J.J.: Die Bewegungsgruppen der Kristallographie. 2. Ausgabe, Birkhäuser, Basel (1966)

Burc3 Burckhardt, J.J.: Zur Geschichte der Entdeckung der 230 Raumgruppen. Arch. Hist. Exact. Sci. 4 (1967/68) 235-246

Burc4 Burckhardt, J.J.: Die Entdeckung der 32 Kristallklassen durch M.L. Frankenheim im Jahre 1826. N. Jb. Miner. Mh. 11 (1984) 481-482

Burk1 Burke, J.G.: Origins of the science of crystals. University of California press, Berkeley and Los Angeles (1966)

Burz1 Burzlaff, H., Fischer, W., Hellner, E. ,Niggli, A.: Zur Entwicklung des Begriffs "Gitterkomplex". Z. Kristallogr. 139 (1974) 246-251

Burz2 Burzlaff, H., Zimmermann, H.: On the choice of origins in the description of space groups. Z. Kristallogr. 152 (1980) 151-179

Buse1 Buser, P.: A geometric approch to invariant subspaces of orthogonal matrices. American Math. Monthly 89(1982) 751

Conw1 Conway, J.H., Sloane, N.J.A.: Laminated lattices. Ann. Math. 116 (1982) 593-620

Coxe1 Coxeter, H.S.M.: Regular polytopes. 3.edition, Collier -Macmillen, Dover, New York (1973)

Curi1 Curie, P.: Sur les répétitions et la symétrie. Comptes rendus Acad. Sci. Paris 100 (1885) 1393-1396

Dade1 Dade, E.C.: The maximal finite groups of 4x4 matrices. Ill. J. Math. 9 (1965) 99-122

Daut1 Dauter, Z.: Domains in pseudocubic $P2_12_12_1$ space group. Abstacts of the Paul Niggli Symposium, Zürich (1984) 23-24

Debr1 Debrunner, H.E.: Zerlegungsrelationen zwischen regulären Polyedern des E^d. Arch. Math. 30 (1978) 656-660

Debr2 Debrunner, H.E.: Ueber Zerlegungsgleichheit von Pflasterpolyedern mit Würfeln. Arch. Math. 35 (1980) 583-587

Dehn1 Dehn, M.: Ueber den Rauminhalt. Math. Ann. 55 (1901) 465-478

Dela1 Delafosse, G.: Recherches sur la cristallisation considérée sous les rapports physiques et mathématiques. Comptes rendus Acad. Sc. Paris, 11 (1840) 394-400

Delo1 Delaunay, B.N.: Sur la partition regulière de l'espace à 4 dimension. Izv. Akad. Nauk. SSSR Otdel. Fiz.-Mat. Nauk. (1929) 79-110, ibid. 145-164

Delo2 Delaunay, B.N.: Neuere Darstellung der geometrischen Kristallographie. Z. Kristallogr. 84 (1932) 109-149. Erratum: ibid. 85 (1933) 392

Delo3 Delone (Delaunay), B.N. and Sandakova, N.N.: Teorija stereoedrow. Trudy. Math. Inst. Akad. Nauk. SSSR, W. A. Steklow 64 (1961)

Delo4 Delone (Delaunay), B.N., Dobilin, N.P., Štogrin, M.I., Galiulin, R.V.: A local criterion for regularity of a system of points. Soviet Math. Dokl. 17 (1976) 319-322

Diri1 Dirichlet, G.L.: Ueber die Reduction der positiven quadratischen Formen mit drei unbestimmten ganzen Zahlen. J. reine angew. Math. 40 (1850) 209-227

DuVa1 Du Val, P.: Homographies, Quaternions and Rotations. Oxford University Press, London (1964)

Eise1 Eisenstein, G.: Tabelle der reducierten positiven quadratischen Formen nebst den Resultaten neuer Forschungen über diese Formen, insbesondere Berücksichtigung auf ihre tabellarische Berechnung. J. reine angew. Math. 41 (1854) 141-190

Enge1 Engel, P.: Ueber Wirkungsbereichsteilungen von kubischer Symmetrie. Z. Kristallogr. 154 (1981) 199-215

Enge2 Engel, P.: Ueber Wirkungsbereichsteilungen von kubischer Symmetrie. II. Die Typen von Wirkungsbereichspolyedern in den symmorphen kubischen Raumgruppen. Z. Kristallogr. 157 (1981) 259-275

Enge3 Engel, P.: On the enumeration of polyhedra. Discrete Math. 41 (1982) 215-218

Enge4 Engel, P., Matsumoto, T., Steinmann, G., Wondratschek, H.: The non-characteristic orbits of the space groups. Z. Kristallogr. Supplement issue (1984)

Enge5 Engel P.: DIRICHLET, a programm to calculate Dirichlet domains of crystallographic orbits. Univ. Bern (1983)

Enge6 Engel P.: SEEBER, a program to calculate the Seeber-reduced form. Univ. Bern (1980)

Enge7 Engel P.: SUB3, a programm to calculate the subgroups of space groups. Univ. Bern (1985)

Enge8 Engel P.: A list of maximal parallelotopes in five-dimensional Euclidean space. Univ. Bern (1985)

Erns1 Ernst, B.: Der Zauberspiegel des M.C. Escher. Verlag Ernst und Moos, München (1978)

Eule1 Euler, L.: Elementa doctrinae solidorum. Opera omnia series prima, Vol. 26, Orell Füssli, Zürich (1953) 71-73. Demonstratio nonnullarum insignium propietatum quibus solide hedris planis inclusa sunt praedita. ibid. 94-108

Fedo1 von Fedorov, E.S.: An Introduction to the Theory of Figures (in Russian). Zipiski imperator St. Petersburg Miner. Obscestra [2], 21 (1885) 1-279; Reprinted: Leningrad (1953)

Fedo2 von Fedorov, E.S.: Simmetrija prawilnych sistem figur (The symmetry of regular systems of figures). Zapiski imperator St. Petersburg Miner. Obscestra [2] 28 (1891) 1-146. English translation by D. and K. Harker: ACA Monograph number 7, Amer. Crystallogr. Ass. (1971)

Fedo3 von Fedorov, E.S.: Zusammenstellung der kristallographischen Resultate des Herrn Schönflies und der meinigen. Z. Kristallogr. 20 (1892) 25-75

Fedo4 von Fedorov, E.S.: Simmetrija na ploskosti (Symmetry in the plane). Zapiski imperator St. Petersburg Miner. Obscestra [2] 28 (1891) 345-390

Fedo5 von Fedorov, E.S.: Theorie der Krystallstructur. Z. Kristallogr. 25 (1896) 113-224

Feje1 Fejes Tóth, J.: Ueber einen geometrischen Satz. Math. Z. 46 (1940) 79-83

Fisc1 Fischer, W.: Existenzbedingungen homogener Kugelpackungen in Raumgruppen tetragonaler Symmetrie. Z. Kristallogr. 133 (1971) 18-42. Existenzbedingungen homogener Kugelpackungen zu kubischen Gitterkomplexen mit weniger als drei Freiheitsgraden. ibid. 138 (1973) 129-146. Existenzbedingungen homogener Kugelpackungen zu kubischen Gitterkomplexen mit drei Freiheitsgraden. ibid. 140 (1974) 50-74

Fisc2 Fischer, W.: Eigenschaften der Heesch-Laves Packung und ihres Kugelpackungstyps. Z. Kristallogr. 143 (1976) 140-155

Fran1 Frankenheim, M.L.: Crystallonomische Aufsätze. ISIS enzyklopädische Zeitung von Oken 5 (1826) 497-515. ibid. 6 (1826) 542-565

Fran2 Frankenheim, M.L.: System der Crystalle. Nova Acta Acad. Caesarea Leopoldino-Carolinae Naturae Curiosorum, 19 (1842) 471-660

Frob1 Frobenius, F.G.: Ueber die unzerlegbaren diskreten Bewegungsgruppen. Sitzungsber. kgl. Preuss. Akad. der Wiss. in Berlin, Phys. Math. Klasse (1911) 654-665

Fuji1 Fuji Television Gallery Co. Ltd.: M.C. Escher's Universe of Mind Play. Odakyo, Japan (1983)

Gado1 Gadolin, A.: Abhandlung über die Herleitung aller krystallographischer Systeme mit ihren Unterabteilungen aus einem einzigen Prinzipe. Ostwald's Klassiker der exakten Wissenschaften Bd. 75, Verlag Engelmann, Leipzig (1896)

Gali1 Galiulin, R.V.: Classification of directions in crystallographic point groups according to the symmetry principle. Acta Crystallogr. A36 (1980) 864-869

Gali2 Galiulin, R.V.: Axiomatic construction of the geometrical principles of crystallography. Sov. Phys. Crystallogr. 24 (1979) 381-383

Gali3 Galulin, R.V.: Delaunay systems. Sov. Phy. Crystallogr. 25 (1980) 517-520

Galo1 Galois, E.: Sur la Théorie des Nombres. Bull. des Sci. math. de Ferussac (1830). Reprint: Galois E.: Ecrits et Mémoires mathémathiques d'Evariste Galois. Paris (1962)

Gaus1 Gauss, C.F.: Recension der "Untersuchungen über Eigenschaften der positiven ternären quadratischen Formen von Ludwig August Seeber". J. reine angew. Math. 20 (1840) 312-320

Gour1 Goursat, E.: Sur les substitutions orthogonal et les divisions regulières de l'espace. Ann. Scient. de l'école normale supérieure 6 (1889) 9-102

Groe1 Groemer, H.: Ueber Würfel- und Raumzerlegungen. El. Math. 19 (1964) 25-27

Grot1 Groth, P.: Elemente der physikalischen und chemischen Krystallographie. R. Oldenbourg, München, Berlin (1921)

Grot2 Groth, P.: Entwicklungsgeschichte der mineralogischen Wissenschaften. J. Springer, Berlin (1926). Reprint: M. Sändig, Wiesbaden (1970)

Grün1 Grünbaum, B., Shephard, G.C.: Isohedral tilings of the plane by polygons. Comm. Math. Helv. 53 (1978) 542-571

Grün2 Grünbaum, B., Shephard, G.C.: Tilings with congruent tiles. Bul. Amer. Math. Soc. 3 (1980) 951-973

Grün3 Grünbaum, B., Shephard, G.C.: Tilings and patterns. Freemann and Co., San Francisco (to appear)

Gubl1 Gubler, M.: Ueber die Symmetrien der Symmetriegruppen: Automorphismengruppen, Normalisatorgruppen und characteristische Untergruppen von Symmmetriegruppen, insbesondere der kristallographischen Punkt- und Raumgruppen. Dissertation ETH. Zürich (1982)

Hadw1 Hadwiger, H.: Vorlesung über Inhalt, Oberfläche und Isoperimetrie. Springer, Berlin (1957)

Hahn1 Hahn, T.: International Tables for crystallography Vol. A. D. Reidel, Dordrecht, Boston (1983)

Hark1 Harker, D.: A Table of the colored crystallographic and icosahedral point groups. Acta Crystallogr. A32 (1976) 133-139

Hark2 Harker, D.: The three-colored three-dimensional space groups. Acta Crystallogr. A37 (1981) 286-292

Hees1 Heesch, H.: Zur Strukturtheorie der ebenen Symmetriegruppen. Z. Kristallogr. 71 (1929) 95-102

Hees2 Heesch, H.: Ueber vierdimensionale Gruppen des dreidimensionalen Raumes. Z. Kristallogr. 73 (1930) 325-345

Hees3 Heesch, H., Laves, F.: Ueber dünne Kugelpackungen. Z. Kristallogr. 85 (1933) 443-453
Herm1 Hermann, C.: Zur systematischen Strukturtheorie III. Ketten und Netzgruppen. Z. Kristallogr. 69 (1929) 250-270
Herm2 Hermann, C.: Zur systematischen Strukturtheorie. IV. Untergruppen. Z. Kristallogr. 69 (1929) 533-555
Herm3 Hermann, C.: Kristallographie in Räumen beliebiger Dimensionszahl. I. Die Symmetrieoperationen. Acta Crystallogr. 2 (1949) 139-145
Hess1 Hessel, J.F.C.: Krystallometrie. Gehler's Phys. Wörterbuch Bd. 5, Leipzig (1830)
Hess2 Hessel, J.F.C.: Krystallometrie oder Krystallonomie und Krystallographie. Ostwald's Klassiker der exakten Wissenschaften Bd. 88/89, Verlag Engelmann, Leibzig (1897)
Hilb1 Hilbert, D.: Mathematische Probleme. Nachr. kgl. Ges. Wiss. Göttingen, Math. Phys. Kl. (1900), 253-297
Hilb2 Hibert, D., Cohn-Vossen, S.: Anschauliche Geometrie. (1931). Reprint: Wiss. Buchgesellschaft, Darmstadt (1973). Englisch translation: Chelsea, New York (1953).
Hill1 Hiller, H.: The crystallographic restriction in higher dimensions. Acta Crystallogr. A41 (1985) 541-544
Hirs1 Hirshfeld, F.L.: Symmetry in the generation of trial structures Acta Crystallogr. A24 (1968) 301-311
Hurl1 Hurley, A.C.: Finite rotation groups and crystal classes in four dimensions. Proc. Camb. Phil. Soc. 47 (1951) 650-661
Hurl2 Hurley, A.C., Neubüser, J. and Wondratschek, H.: Crystal classes of four-dimensional space R^4. Acta Crystallogr. 22 (1967) 605
Igle1 Iglesias, J.E.: A formula for the number of closest packings of equal spheres having a given repeat period. Z. Kristallogr. 155 (1981) 121-127
Jarr1 Jarrat, J.D., Schwarzenberger, R.L.E.: Coloured plane groups. Acta Crystallogr. A36 (1980) 884-888
Jess1 Jessen, B.: The algebra of polyhedra and the Dehn-Sydler theorem. Math. Scand. 22 (1968) 241-256
Jone1 Jones, O.: Grammer of ornament. Bernard Quatrich, London (1868)
Jord1 Jordan, C.: Sur les groupes de mouvements. Comptes rendus Acad. Sci. Paris 65 (1867) 229-232
Koch1 Koch, E.: Die Wirkungsbereichspolyeder und Wirkungsbereichsteilungen zu kubischen Gitterkomplexen mit weniger als drei Freiheitsgraden. Z. Kristallogr. 138 (1973) 196-215
Koch2 Koch, E., Fischer, W.: Zur Bestimmung asymmetrischer Einheiten kubischer Raumgruppen mit Hilfe von Wirkungsbereichen. Acta Crystallogr. A30 (1974) 490-496

Koch3 Koch, E., Fischer W.: Automorphismengruppen von Raumgruppen und die Zuordnung von Punktlagen zu Konfigurationslagen. Acta Crystallogr. A31 (1975) 88-95

Kopt1 Koptsik, V.A.: Advances in theoretical crystallography. Color symmetry of defect crystals. Kristall und Technik 10 (1975) 231-245

Křiv1 Křivý S. and Gruber B.: A unified algorithm for determining the reduced (Niggli) cell. Acta Crystallogr. A32 (976) 297-298

Lagr1 Lagrange, J.L.: Recherches d'arithmetique. Oeuvre III (1773) 695-795

Lave1 Laves, F.: Ebenenteilung in Wirkungsbereiche. Z. Kristallogr. 76 (1931) 277-284

Lave2 Laves, F.: Ebenenteilung und Koordinationszahl. Z. Kristallogr. 78 (1931) 208-241

Lawr1 Lawrenson, J.E., Wondratschek, H.: The extraordinary orbits of the 17 plane groups. Z. Kristallogr. 143 (1976) 471-484

Leec1 Leech, J., Sloane, N.J.A.: Sphere packings and error-correction codes. Can. J. Math. 23 (1971) 718-745

Lenz1 Lenz, H.O.: Mineralogie der alten Griechen und Römer. M. Sändig, Wiesbaden (1861). Reprint: M. Sändig, Wiesbaden (1966)

Löck1 Löckenhoff, H.D., Hellner, E.: Die Wirkungsbereiche der invarianten kubischen Gitterkomplexe. N. Jb. Miner. Mh. (1971) 155-174

Mack1 Mackay, A.L., Pawley, G.S.: Bravais lattices in four-dimensional space. Acta Crystallogr. 16 (1963) 11-19

Marx1 Marx, C.M.: Geschichte der Kristallkunde. M. Sändig, Wiesbaden (1825). Reprint: M. Sändig, Wiesbaden (1970)

Mats1 Matsumoto, T., Nowacki, W.: On densest packing of ellipsoids. Z. Kristallogr. 123 (1966) 401-421

McLa1 McLarnan, T.J.: The numbers of polytopes in close-packings and related structures. Z. Kristallogr. 155 (1981) 269-291

McMu1 McMullen, P.: Convex bodies which tile space by translation. Mathematica 27 (1980) 113-121

Mich1 Michel, L.,Mozrzymas, J.: Structure des classes de conjugaison d'un groupe cristallographique. Comptes randus Acad. Sc. Paris, série II, 299 (1984) 387-390

Mill1 Miller, W.H.: A treatise on crystallography. Cambridge (1839)

Mink1 Minkowski, H.: Diskontinuitätsbereich für arithmetische Aequivalenz. J. reine angew. Math. 129 (1905) 220-274

Naum1 Naumann, C.F.: Lehrbuch der reinen und angewandten Krystallographie. Leipzig (1830)

Nef1 Nef, W.: Lehrbuch der linearen Algebra. Birkhäuser, Basel (1966) English translation by J.C. Ault: Mc Graw-Hill (1967)

Nef2 Nef, W.: Beiträge zur Theorie der Polyeder. Herbert Lang, Bern (1978)

Neub1 Neubüser, J. Wondratschek, H. and Bülow, R.: On crystallography in higher dimensions. I. General definitions. Acta Crystallogr. A27 (1971) 517-520

Neub2 Neubüser, J., Plesken, W. and Wondratschek, H.: An emendatory discursion on defining crystal systems. Match 10 (1981) 77-96

Neub3 Neubüser, J., Wondratschek, H.: Untergruppen der Raumgruppen. Kristall und Technik 36 (1966) 529-543

Nigg1 Niggli, P.: Geometrische Kristallographie des Diskontinuums. Bornträger, Leipzig (1918)

Nigg2 Niggli, P.: Die regelmässige Punktverteilung längs einer Geraden in einer Ebene. Z. Kristallogr. 63 (1926) 255-274

Nigg3 Niggli, P.: Handbuch der Experimentalphysik: Kristallographische und Strukturtheoretische Grundbegriffe. Akad. Verlagsgemeinschaft, Leibzig (1928)

Nigg4 Niggli, P., Nowacki, W.: Der arithmetische Begriff der Kristallklassen und die darauf fussende Ableitung der Raumgruppen. Z. Kristallogr. 91 (1935) 159-184

Nigg5 Niggli, P.: Die topologische Strukturanalyse. I. Z. Kristallogr. 65 (1927) 391-415, II. ibid. 68 (1928) 404-466

Nowa1 Nowacki, W.: Die nichtkristallographischen Punktgruppen. Z. Kristallogr. 86 (1933) 19-31

Nowa2 Nowacki, W.: Die euklidischen dreidimensionalen, geschlossenen und offenen Raumformen. Comm. Math. Helv. 7 (1934) 81-93

Nowa3 Nowacki, W.: Homogene Raumteilung und Kristallstruktur. Diss. ETH Zürich (1935)

Nowa4 Nowacki, W.: Space groups not always derivable by parallelohedra and subdivision into stereohedra. Science 174 (1971) 52

Nowa5 Nowacki, W.: Bemerkungen zur Geschichte der Raumgruppen Symbole von Fedorov, Schönflies und Hermann-Mauguin. Z. Kristallogr. 135 (1972) 145-158

Nowa6 Nowacki, W.: Symmetrie und physikalisch-chemische Eigenschaften kristallisierter Verbindungen. V. Ueber Ellipsenpackungen in der Kristallebene. Schweiz. Miner. Petr. Mitt. 28 (1948) 502-508

Patt1 Patterson, A.L., Love; W.E.: Remarks on the Delaunay reduction. Acta Crystallogr. 10 (1957) 111-116

Ples1 Plesken, W. and Pohst, M.: On maximal finite irreducible subgroups of GL(n,Z). I. The five and seven dimensional case. Math.Comp. 31 (1977) 536-551. II. The six dimensional case. ibid. 31 (9177) 536-573. III. The nine dimensional case. ibid. 34 (1980) 245-258. IV. Remarks on even dimension with applications to n=8. ibid. 34 (1980) 259-275. V. The eight dimensional case and complete description of dimensions less than ten. ibid. 34 (1980) 277-301.

Ples2 Plesken, W.: Bravais groups in low dimensions. Match 10 (1981) 97-119

Ples3 Plesken, W., Hanrath W.: The lattices of six-dimensional euclidean space. Math. Comp. 43 (1984) 573-587
Ples4 Plesken, W.: Personal communication (1985)
Poly1 Polyá, G.: Ueber die Analogie der Kristallsymmetrie in der Ebene. Z. Kristallogr. 60 (1924) 278-282
Rein1 Reinhardt, K.: Ueber die Zerlegung der Ebene in Polygone. Diss. Univ. Frankfurt (1918)
Rema1 Remak, R.: Ueber die Minkowskische Reduktion. Composito math. 5 (1938) 368
Roge1 Rogers, C.A.: Packing and Covering. Cambridge University Press, (1964)
Rome1 Romé de l'Isle, J.B.L.: Crystallographie, ou description des formes propres a tous les corps du régne minéral. 4 volumes. Paris (1783)
Rysk1 Ryškov, S. S.: On maximal finite groups of integer nxn matrices. Sov. Math. Dokl. 13 (1972) 720-724
Sah1 Sah, C.-A.: Hilbert's third problem: scissors congruence. Research Notes in Mathematics, Vol. 33. Pitman, San Francisco (1979)
Schl1 Schläfli, J.: Theorie der vielflachen Kontinuität. Gesammelte mathematische Abhandlungen, Vol. 1, Birkhäuser, Basel (1950) 189-191
Schl2 Schläfli, J.: Ueber invariante Elemente einer orthogonalen Substitution, wenn dieselbe als Ausdruck einer Bewegung jeder Gruppe von Werthen der Variablen aus dem identischen Zustande in den transformierten gefasst wird. J. reine angew. Math. 65 (1866) 185-187. Gesammelte mathematische Abhandlungen, Vol. 2, Birkhäuser, Basel (1953) 365-367
Schö1 Schönflies, A.: Ueber Gruppen von Bewegungen. Math. Ann. 28 (1987) 319-342
Schö2 Schönflies, A.: Ueber reguläre Gebietsteilungen des Raumes. Nachrichten von der kgl. Gesell. der Wiss. und der Georg-Augusts Univ. zu Göttingen 9 (1888) 223-237
Schö3 Schönflies, A.: Krystallsysteme und Krystallstructur. B. G. Teubner, Leipzig (1891). Reprint: Springer, Berlin (1984)
Shub1 Shubnikov, A.V. and Belov, N.V.: Colored Symmetry. Pergamon Press, Oxford (1964)
Schu1 Schulte, E.: Tiling three-space by combinatorially equivalent convex polytopes. Proc. London Math. Soc. 49 (1984) 128-140
Schw1 Schwarzenberger, R.L.E.: N-dimensional crystallography. Research Notes in Mathematics, Vol. 41. Pitman, San Francisco (1980)
Schw2 Schwarzenberger, R.L.E.: Decomposition of crystal families and graphical representations of space groups and asymptotic estimates for large dimensions. Match 10 (1981) 121-122
Seeb1 Seeber, L.A.: Versuch einer Erklärung des innern Baues der festen Körper. Gilbert's Ann. der Physik 76 (1824) 229-248

Seeb2 Seeber, L.A.: Untersuchungen über die Eigenschaften der positiven ternären quadratischen Formen. Freiburg (1831)

Sell1 Selling, E.: Ueber die binären und ternären quadratischen Formen. J. reine angew. Math. 77 (1874) 143-229

Sene1 Senechal, M.: Point groups and color symmetrie. Z. Kristallogr. 142 (1975) 1-25

Sene2 Senechal, M.: A simple characterization of the subgroups of space groups. Acta Crystallogr. A36 (1980) 845-850

Sene3 Senechal, M.: Morphisms of crystallographic groups: Kernels and images. J. Math. Phys. 26 (1985) 219-228

Sino1 Sinogowitz, U.: Herleitung aller homogenen nicht kubischen Kugelpackungen. Z. Kristallogr. 105 (1943) 23-52

Sohn1 Sohncke, L.: Die Gruppierung der Molecüle in den Krystallen. Poggendorffs Ann. der Physik 132 (1867) 75-106

Sohn2 Sohncke, L.: Die regelmässigen ebenen Punktsysteme von unbegrenzter Ausdehnung. J. reine angew. Math. 77 (1874) 47-101

Sohn3 Sohncke, L.: Entwicklung einer Theorie der Krystallstructur. B.G. Teubner, Leibzig (1879)

Sohn4 Sohncke, L.: Erweiterung der Theorie der Krystallstructur. Z. Kristallogr. 14 (1888) 426-446

Sohn5 Sohncke, L.: Die Entdeckung des Einteilungsprincips der Krystalle durch J.F.C. Hessel. Z. Kristallogr. 18 (1890) 486-498

Spei1 Speiser, A.: Die Theorie der Gruppen endlicher Ordnung. 4. edition, Birkhäuser, Basel (1956)

Sten1 Stenonis, N.: De Solido intra Solidum naturaliter contento Dissertationis Prodromus. Florentiae (1669). Translated in Niels Stenson: Das Feste im Festen. Ostwalds Klassiker der exakten Wissenschaften Bd 3, Akad. Verlagsgesellschaft, Frankfurt (1967)

Štog1 Štogrin, M.I.: Regular Dirichlet-Voronoi partitions for the second triclinic group. Proceedings of the Steklov Institute of Mathematics 123 (1973). English translation by E.J.F. Primrose: Amer. Math. Soc. (1975)

Sydl1 Sydler J.-P.: Conditions nécessaires et suffisantes pour l'équivalence des polyèdres de l'espace euclidean à trois dimensions. Comment. Math. Helv. 40 (1965) 43-80

Venk1 Venkov, B.A.: On a class of euclidean polytopes (in Russian). Vestnik Leningrad Univ. (Ser. Mat. Fiz. Him.) 9 (1954) 11-31

Voro1 Voronoï, M.G.: Nouvelles applications des paramètre continus à la théorie des formes quadratiques. J. reine angew. Math. 134 (1908) 198-287. ibid. 134 (1909)

Waer1 Van der Waerden, B.L., Burckhardt, J.J.: Farbgruppen. Z. Kristallogr. 115 (1961) 231-234

Waer2 Van der Waerden, B.L., Gross, H.: Studien zur Theorie der quadratischen Formen. Birkhäuser, Basel (1968)

Webe1 Weber, L.: Die Symmetrie homogener ebener Punktsysteme. Z. Kristallogr. 70 (1929) 309-327

Webe2 Weber, O: Ueber die Reduktion und die Darstellung positiver quaternärer quadratischer Formen. Comm. Math. Helv. 36 (1962) 181-213

Weig1 Weigel, D., Phan, Th., Veysseyre, R.: Sur les symboles géomètriques des groupes ponctuels cristallographiques de l'espace euclidien à 4 dimensions. Comptes rendus Acad. Sc. Paris, série II, 298 (1984) 825-828, ibid. 300 (1985) 51-54

Weis1 Weiss, C.S.: Uebersichtliche Darstellung der verschiedenen natürlichen Abtheilungen der Krystallsysteme. Abh. der kgl. Akad. der Wiss. in Berlin (1814/15) 289-336

Weis2 Weiss, C. S.: Ueber eine verbesserte Methode für die Bezeichnung der verschiedenen Flächen eines Krystallsystems. Abh. der kgl. Akad. der Wiss. in Berlin (1816/17) 286-336

Weis3 Weissenberg, K.: Aufbau der Kristalle. Z. Kristallogr. 62 (1925) 13-51, ibid. 612, ibid. 63 (1926) 221

Whit1 Whittaker, E.J.W.: Graphic representation and nomenclature of four-dimensional crystal classes. III. A notation for the crystal classes. Acta Crystallogr. A40 (1984) 404-410

Wien1 Wiener, Chr.: Grundzüge der Weltordnung. Leipzig, Heidelberg (1863)

Wiet1 Wieting, T.: The mathematical theory of chromatic plane ornaments. Dekker, (1982)

Wond1 Wondratschek, H., Bülow, R., Neubüser, J.: On crystallography in higher dimensions. III. Results in R^4. Acta Crystallogr. A27 (1971) 523-535

Wond2 Wondratschek H.: Extraordinary orbits of space groups. Theoretical considerations. Z. Kristallogr. 143 (1976) 460-470

Wulf1 Wulff L.: Ueber die regelmässigen Punktsysteme. Z. Kristallogr. 13 (1887) 503-566

Zamo1 Zamorzaev, A.M.: Generalization of the space groups. Dissertation, Leningrad (1953)

Zamo2 Zamorzaev, A.M., Tsekinovskii, B.V.: Four-dimensional Bravais lattices. Sov. Phys. Crystallogr. 13 (1968) 165-168

Zamo3 Zamorzaev, A.M., Palistrant, A.F.: Antisymmetry, its generalizations and geometrical applications. Z. Kristallogr. 151 (1980) 231-248

Zass1 Zassenhaus, H.: Ueber einen Algorithmus zur Bestimmung der Raumgruppen. Comm. Math. Helv. 21 (1948) 117-141

Subject Index

A

abstract group 9, 91, 119
adjacency scheme 208
Agricola 10
Alexander 197
algorithm
  of Delaunay 60-61
  of Gauss 29, 32, 82, 87
  of Křivý-Gruber 51
  of Seeber 48-59, 67
Aristotle 237
arithmetic
  Bravais class 124
  crystal class 123, 141-143, 150, 153, 158, 161
Ascher 200
asymmetric unit 151
axiom 1

B

Baranovskii 223
Barlow 199, 248
Belov 149, 200
belt 38, 39, 40
Bieberbach 62, 64, 199
  theorem 22-25
Blichfeldt 65, 248
Bohm 195
Boolean union 20
Bravais 121, 149, 238
  class
    arithmetic 124
    geometric 124, 128
  point group 124-141
  type of lattice 124, 128, 135, 141
Brillouin zone 13
Brown 122, 164, 165, 200
Buerger 150
Bülow 111, 143, 149, 200
Burckhardt 120, 150, 199
Buser 72

C

characteristic
  Bravais point group 142
  crystal form 144, 146
  orbit 184
  polynom 69, 71, 75
chirality 6, 91, 199
classification 10, 113, 119, 123, 203, 208, 241
coboundary 156
cocycle 156
cohomology group 156
colour group 191-195
colouring 192
combinatorial type 34, 39, 203
computer program 21, 50, 181
construction
  of Euler 95
  of Sohncke 12, 189
contragredient 26, 30, 91
convex polytope 202
coordinate system 4, 5
  cartesian 4, 6, 19, 27, 62, 70, 73, 130
  crystal 4
corona 225
correspondence rule 33, 144
covering property 16, 42
Coxeter 107, 110, 121, 236
crystal
  class
    arithmetic 123, 141, 150, 153
    geometric 94, 101
    maximal 111
  facet 144
  family 116-118
  form 144-149
    characteristic 144
    closed 144
    open 144
  system 52-59, 114, 115, 122
crystallographic

group 9
orbit 183-190, 242
point group 89
restriction 71, 119
rotation 77, 78, 83, 94
Curie 11, 199

D

d-face 16, 18, 19, 204
d-flag 204
d-subordination symbol 203
d-superordination symbol 204
Dade 111
Dauter 238
Debrunner 236, 239
Dehn 234, 239
function 235
Delafosse 11
Delaunay 1, 12, 35, 59, 66, 222, 238, 239
tetrahedron 60
Dirichlet 13, 48, 66, 238
domain 11-21, 152
partition 16, 22, 23, 209-221
parallelotope 34
discontinuum 1
dissection 234-237
dual
basis 4
polyhedron 96
polytope 108-110

E

eigenvalue 69
eigenvector 69
Eisenstein 66
enantiomorph 6, 143
equivalence
class 10
of symmetry groups 91
relation 10, 27, 44, 114
Escher 200
Euclidean
group 68
space 1, 3-5, 34
Euler
construction 95
relation 204, 210
totient function 75, 76
exact sequence 155
extent 43
extreme point 15

F

face-to-face 36, 201
facet 14, 15, 18
vector 32, 35-43, 144
Fedorov 12, 39, 143, 150, 164, 199, 238
Fischer 184, 246, 248
Frankenheim 120, 141, 149, 238
free abelian group 153
Frobenius 5, 199
congruence 155
subgroup congruence 171
symbol 5
fundamental region 151, 152

G

Gadolin 121, 122
Galiulin 149, 165, 239
Galois 120
gamma function 65, 241
Gauss 29, 66, 248
generating set 23, 92, 243
geometric
Bravais class 124, 128
class 91
crystal class 94, 101, 110, 111
extension 226
Gessner 10
glide reflection 86
Goursat 121
Groemer 239
Groth 150
group 8
Euclidean 225
group extension 155
Gruber 50, 67
Grünbaum 201, 243
Guglielmini 237

H

Hadwiger 239
half-space 13
intersection 13, 19, 21, 202
Harker 193
Haüy 10, 150
Heesch 200, 246
Hellner 238
Hermann 75, 76, 101, 121, 162, 170, 197
Hessel 120
hierarchical structure 17-19
Hilbert 1, 12, 199, 234, 239
holohedry 113, 122
homogeneous 240
honycombe 34
Hook 248
Hurley 122
Huygens 10, 248

I

interstitial ball 1, 45, 46
involution 102-118, 126, 128
Isle 10
isohedral 201
isometry 6
proper 6
isomorphism type 119
isotropy group 9

J

Janner 200
Jarrat 193
Jessen 236
Jordan 198

K

Kepler 10, 248
Klein 199
Koch 184, 238
Koptsik 192
Křivý 50, 67

Kronecker product 133

L

Lagrange 44, 66
reduction 45
lattice 22-43
basis 25-27, 42, 43, 44-65, 123-141
hyperplane 32
packing 240, 246-247
plane 32-33, 80, 144
symmetry 123-141
Laue 238
Laue class 173
Laves 210, 238, 246
law of rationality 32, 120, 149
Lawrenson 188
Leech 247, 248
locally finite 202
Löckenhoff 238

M

Mackay 149
Matsumoto 188, 246
Mauguin 101, 162
maximal
stereotope 211
McMullen 40
metric tensor 4, 6, 25, 27, 28, 44, 48
Michel 161
Miller indices 33
minimal
Bravais point group 142
group 167
Minkowski 44, 63, 65, 66, 248
reduction 62-65
monohedral 201
mosaic 195

N

Naumann 150
Nef 17, 82
Neubüser 122, 200

Niggli 2, 12, 47, 50, 66, 122, 143, 150, 196, 200, 238, 242, 248
normal 201
normal subgroup 97, 113
normalizer 165, 169
affine 128, 131, 158, 165, 171, 185, 192
Euclidean 167, 189, 216
orthogonal 142, 165
Nowacki 121, 143, 150, 190, 238, 243
numbers
of belts 40, 41
of Bravais classes
arithmetic 141
geometric 128
of Bravais types of lattices 141
of colour-group types 194, 196
of crystal classes
arithmetic 143
geometric 111
of facets 36
of maximal matrix groups 111
of non-isomorphic groups 120
of orbit types 189
of polyhedra types 206
of space group types 165
of subperiodic groups 198
of types of crystal forms 145
of Wyckoff sets 184

O

orbit 42, 151, 242
characteristic 184
topological type 185, 221
orthogonal
basis 27-32
group 142
relation 6

P

packing
of balls 240-247
of ellipses 243
parallelogon 34, 45, 48
parallelohedron 39, 46
parallelotope 33-43, 64, 221-224
belt 38, 39, 40
primitive 36
zone 38
Parker 148, 150
partial order 17
Patterson 61
perfect colouring 193
plane group 161-164
Platonic solids 96, 237
Plesken 111, 150, 161
point group 9, 22
crystallographic 89
determination 100
first kind 92, 93, 96, 97, 114
in E^2 93-94
in E^3 94-102
in E^n 102-111
point lattice 11, 25, 33, 123
apparent 190
Polyá 199
polyhedron 17
polytope 15, 18
scheme 205
principal
plane 72, 80, 102-118, 132
point 190
point group 158, 190
projection operator 79, 134

Q

Q-class 91
Q-region 14, 15, 21, 38
quadratic form 26, 28, 44-65, 124, 132, 134, 135, 149

R

reciprocal basis 4, 32
reduced form 44

reduction 44-65
of Lagrange 45-46
of Minkowski 62-65
of Seeber 46-59
of Selling 59-61
reference point 13
regular
point system 2, 9, 11, 22, 68, 89, 123, 151, 158, 172, 184, 241
polytope 108-110, 236, 247
splitting 157, 158
regularity
condition 225-233
Reinhardt 209, 238
relative
interior point 15
prime 75, 130
Remak 65
representation 9
integral 69, 71
irreducible 74
rigid motion 2, 5-7
Rogers 247, 248
root
class 111-118
system 112, 113-118
vector 102-118, 128
rotation 70, 72, 81
angle 71
axis 78, 84, 95
order 75
part 5, 22
proper 6, 77, 94
rotoinversion 78
rotoreflection 6, 11, 78, 99, 101, 121
(r,R)-system 1, 13, 15, 16, 225
Ryškov 223

S

Sah 234
Schläfli 72, 76, 204
Schönflies 12, 78, 93, 164, 199, 238
notation 78, 93
Schur 62
Schwarzenberger 161, 165, 167, 193
screw
rotation 86, 152
vector 69, 84-88, 152
Seeber 11, 44, 47, 66, 149, 238, 248
reduction 46-59
selection rules 44, 63
Selling 66
reduction 59-61
Senechal 170, 182, 193
Shephard 201, 243
shift vector 5, 69, 84-88, 155
simplex 6, 18, 19, 247
Sinogowitz 246, 248
Sloane 247
Sohncke 1, 2, 11, 12, 188, 198
construction 12, 189
space
group 9, 151-165
type 151
of quadratic forms 132
partition 201-239
Speiser 192, 200
stabilizer 9, 144, 145, 151, 165, 208, 213, 222, 225, 241
star 133
Steinmann 188
Stenson 10
stereotope 202, 207
Štogrin 222, 238
subgroup
class equivalent 170
normal 93, 182
of point group 102
of space group 170
translation equivalent 170
subordination 17, 19, 20
subperiodic group 195
of borders 196
of layers 197
supporting plane 40
Sydler 236, 239
symmetry
operation 7-9, 22
crystallographic 68
general 84
minimax 152
representation of 69
scaffolding 95, 119, 165

point group 90, 92, 98, 142, 152, 165
space group 153
support 81-84, 86, 133, 134
point group 90
space group 183
symmorphic space group 143, 150, 155, 161-164, 218

T

tessellation 207, 210
Toth 243
trace 70, 134
transformation
of basis 25, 69
transitive 9, 22
translation
group 153
part 5
type 10
of colour groups 192
of crystal forms 144, 145
of isomorphic groups 119
of orbits 184
of parallelohedra 39, 221
of parallelotopes 222
of polytope 203
of regular packing 242
of space partitions 208-221
of Wyckoff sets 184
of $\mathbb{Z}$-reduced forms 52-59

U

unimodular 26, 44, 68, 90, 123

V

valence 211
Van der Waerden 63
vector space 4
Venkov 40
Voronoi 34, 35, 36, 66, 211, 222, 238
region 13

W

Weber, L. 197, 200
Weber, O. 63
Weigel 116
weight lattice 172
Weiss 119, 122, 150
Weissenberg 158, 190
Whewell 150
Whittacker 116
Wiener 11
Wieting 193
Wigner-Seitz cell 13
Wondratschek 122, 188, 200
Wulff 11, 199
Wyckoff
position 183
set 183

Z

$\mathbb{Z}$-class 123
$\mathbb{Z}$-irreducible 74-77, 126-132, 161
Zamorzaev 149, 193, 197
Zassenhaus 122, 200
zone 38
reduction 221

GPSR Compliance

The European Union's (EU) General Product Safety Regulation (GPSR) is a set of rules that requires consumer products to be safe and our obligations to ensure this.

If you have any concerns about our products, you can contact us on ProductSafety@springernature.com

In case Publisher is established outside the EU, the EU authorized representative is:

Springer Nature Customer Service Center GmbH
Europaplatz 3
69115 Heidelberg, Germany

Batch number: 10620141

Printed by Printforce, the Netherlands